Signal Processing with Free Software

FOCUS SERIES

Series Editor Francis Castanié

Signal Processing with Free Software

Practical Experiments

François Auger

iSTE

WILEY

First published 2014 in Great Britain and the United States by ISTE Ltd and John Wiley & Sons, Inc.

ISTE Ltd
27-37 St George's Road
London SW19 4EU
UK

www.iste.co.uk

John Wiley & Sons, Inc.
111 River Street
Hoboken, NJ 07030
USA

www.wiley.com

Library of Congress Control Number: 2013955372

British Library Cataloguing-in-Publication Data
A CIP record for this book is available from the British Library
ISSN 2051-2481 (Print)
ISSN 2051-249X (Online)
ISBN 978-1-84821-463-7

FSC www.fsc.org MIX Paper from responsible sources FSC® C013604

Printed and bound in Great Britain by CPI Group (UK) Ltd., Croydon, Surrey CR0 4YY

Table of Contents

Preface

This book is based on the teaching documents used in the Physical Measurements Department at the University Institute of Technology in Saint-Nazaire, France, and is aimed at cooperative work/study engineering schools. It is not intended to provide a full grounding in signal processing for which readers should consult the reference works cited in the bibliography.

I wish to thank my colleagues (M. Belkadi, T. Hairie, F. Poyer, H. Menana, J.-C. Olivier, B. Velay and Z. Shi) who have participated in this teaching activity and helped me to improve the quality of the content, and my students (in particular, A. Baudry, A. Bouvet, F. Gibon, R. Monnier and S. Vaillant) who, through their questions, have helped me to clarify and add certain exercises.

The command files and programs mentioned in this book are available at http://www.univ-nantes.fr/auger-f (under "informations complémentaires"/additional information).

François AUGER
December 2013

Introduction

The aim of these practical exercises is to demonstrate some of the possibilities of practical implementation of signal processing techniques. To do this, we will use free software applications, including SoX, `Audacity` and `Visual Analyser`, focusing on their uses in real situations.

SoX[1] is known as the "Swiss army knife of sound processing". Its main application lies in the handling of acoustic signals, but it may also be used for other types of signals, and nothing prevents its use for signals produced by any sensor (in addition to microphones). SoX is a *free*, which means that it may be used, studied and modified freely by users, and *multiplatform*, which means it may be executed using Windows, Linux or Mac OsX, software package. It is a command line application (see Figure I.1), meaning that it does not include a human–machine interface using menus, buttons and fields, but is used by calling up the program

1 Available for Windows, Linux and Mac OsX at http://sox.sourceforge.net. It was created in 1991 by Lance Norskog. The first version was known as AUX (for *Aural Exchange*). Since 1996, it has been maintained by Chris Bagwel. The program was in version 14.3.0 in January 2010, 14.3.1 in January 2011 and 14.4.0 in March 2012, showing the vitality and maturity of the development.

using parameters for the signals to process and the operations to apply. This method means that SOX may be used in application software written in any language. These command lines systematically take the general form:

```
sox InputFile OutputFile Effect EffectParameters
```

Throughout these practical exercises, SOX may be used with Windows, either from a DOS command window[2], or by creating and running command files[3], using the `.bat` extension. These command files correspond to programs containing sequentially executed instructions.

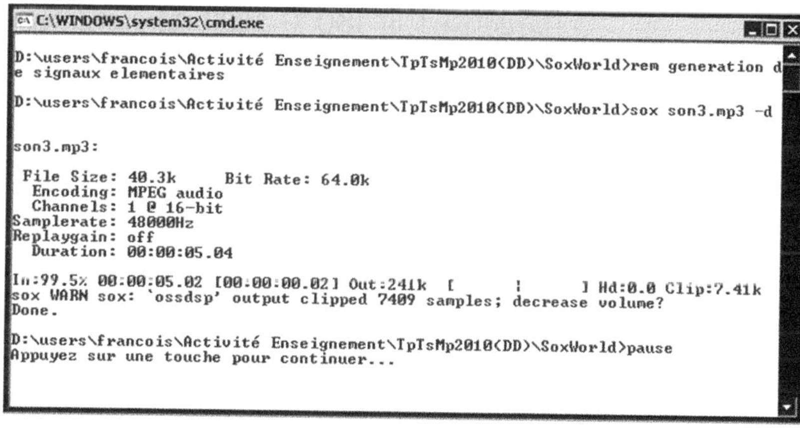

Figure I.1. *Example of the use of* SOX *in a Windows command window*

2 In Windows, Marko Bozicovic's `Console` program, available at http://sourceforge.net/projects/console/, is useful.

3 See http://en.wikipedia.org/wiki/Batchfile. For further information on the commands available using DOS/Windows, see http://en.wikipedia.org/wiki/ListofMS-DOScommands.

The documentation for this program provides a succinct description of its possibilities. The aim of these practical exercises is to present certain signal generation, analysis and treatment functions offered by the program in order to correctly understand and use these possibilities. We will only consider processing operations that are not specific to the field of acoustics; readers who are interested in acoustic and musical effects may wish to carry out further, specific investigations.

Ch is a multiplatform C/C++ interpreter designed for scientific calculations and for teaching algorithmics and computer programming. It conforms to ISO C 1990 and a large part of ISO C 1999. However, it only accepts part of C++. Ch was developed by Harry H. Cheng (University of California, Davis), who passed the responsibility of further development and commercialization to *SoftIntegration Inc.* A "student" version is freely available to users who register with the website http://www.softintegration.com.

Audacity[4] is another free, multiplatform software package. Its functions are sometimes comparable with, but generally complementary to, those offered by SoX: like SoX, it also offers generation, processing and analysis functions for acoustic signals. It operates in a graphical environment (see Figure I.2), enabling easy visualization of signals.

4 Available at http://audacity.sourceforge.net.

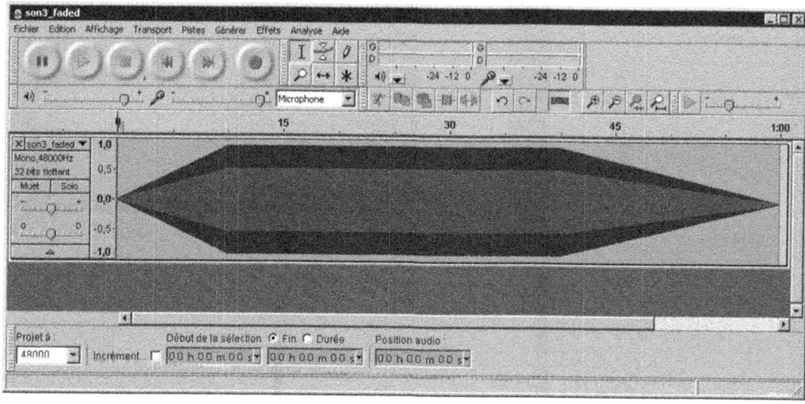

Figure I.2. *Main window of* `Audacity`

Finally, `Visual Analyser` is a free program[5] designed for Windows, which provides an oscilloscope, a spectrum analyzer, a low frequency (LF) generator and a frequency meter (see Figure I.3). It is connected to the personal computer (PC) sound card by default, but may also be used with other acquisition cards.

5 Available at http://www.sillanumsoft.com/.

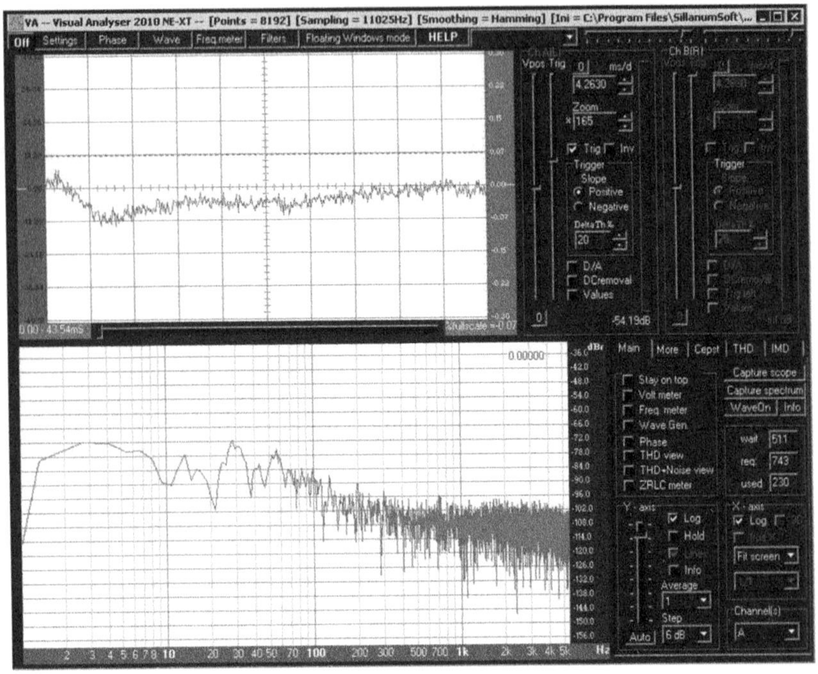

Figure I.3. *Main window of* `Visual Analyser`

1

Generating Elementary Signals

1.1. General points

A signal is a *function* of one or more variables, which has the specificity of being generated by a physical phenomenon and of carrying information rather than energy. Simple examples of signals include speech signals, acoustic signals and the electrical voltages provided by sensors. We may distinguish between different types of signals:

– Continuous time signals, which have a (measurable) value at any time.

– Discrete time signals, which only have (measurable) values at certain moments.

– Sampled signals, which are time-discrete signals for which the time gap between two consecutive measurements is constant.

– Deterministic signals, for which the value may be perfectly predicted using a mathematical model to calculate exact values at any time.

– Random signals, which cannot be precisely predicted, either because it is impossible or because it is not necessary.

– Causal signals, which are null prior to the instant selected as time 0. These signals show the repercussions of a particular event occurring at time 0.

– Stationary signals, which carry the same information throughout the observation period. All periodic signals are stationary, but not all stationary signals are periodic: for example, the deterministic signal defined by the expression $x(t) = 10\sin(2\pi 50t) + 5\sin(2\pi 50\sqrt{2}t)$ is not periodic, as it is the sum of two sinusoids with frequencies in an irrational relationship.

1.2. Generation of elementary wave forms

To test the processing operations and analyses shown later in this book, it is useful to generate synthetic test signals (which are not the result of the observation of a real physical phenomenon). To do this, we use the `synth` command. The command file below (`Generation1.bat`) includes comment lines (which begin with the instruction `rem`) and six calls to the SoX program:

```
rem generation de signaux elementaires avec SoX
rem F. Auger, IUT Saint-Nazaire, dep. MP, dec. 2009

sox -n s1.mp3 synth 3.5 sine 440
sox -n s2.wav synth 90000s sine 660:1000
sox -n s3.mp3 synth 1:20 triangle 440
sox -n s4.mp3 synth 1:20 trapezium 440
sox -V4 -n s5.mp3 synth 6 square 440 0 0 40
sox -n s6.mp3 synth 5 noise

for %%i in (200,300,400) do ^
sox -n s7_%%i.mp3 synth 15 sine %%i

rem ecoute du resultat : le fichier de sortie est le canal
rem de sortie par defaut (-d), c'est a dire la carte son

sox s1.mp3 -d

pause
```

– The first command generates a file `s1.mp3` containing a recording of 3.5 s of a sinusoid with a frequency of 440 Hz. The `-n` (null) command is used to indicate the absence of an input signal.

– The second command generates a file `s2.wav` containing a recording of $90,000$ samples of a sinusoidal signal with a frequency that varies in a linear manner from 660 to $1,000$ Hz.

– The third command generates a file `s3.wav` containing a recording of 1 min and 20 s of a triangular signal.

– The fourth command generates a file `s4.wav` containing a recording of 1 min and 20 s of a trapezoidal signal.

The fifth command generates a file `s5.wav` containing a recording of 6 s of a square signal with a null offset (no continuous component), a null phase and a duty cycle of 50%. The command `-V4` is used to choose the amount of information shown by SoX: its verbosity ranges from 0 (no information sent back to the user) to 4 (maximum information displayed).

– The sixth command generates a file `s6.wav` containing a recording of 5 s of a white noise with a value ranging from 1 to 1 with an average value of 0.

– The seventh command uses a repetition structure to generate three files containing a recording of 15 s of a sinusoid. The variable `%%i` is used both for the file name and for the frequency of the sinusoid. The symbol `^` indicates that the instruction continues onto the following line.

– The final command sends the contents of the file `s1.mp3` to the default peripheral `-d`, i.e. the sound card, allowing us to listen to the signal.

EXERCISE 1.1.– Write a command file to generate 10 files containing recordings of 2.5 s of sinusoids, for which the frequencies should be linearly spaced from 100 to $1,000$ Hz. What is the analytical expression of these signals?

1.3. Elementary signal processing operations

A number of elementary operations may be applied to existing signals. The command file below (called `Generation2.bat`) shows some examples:

```
rem transformations elementaires de signaux avec SoX
rem F. Auger, IUT Saint-Nazaire, dep. MP, jan. 2010

sox s1.mp3 s1_faible1.mp3 vol -6 dB
sox s1.mp3 s1_faible2.mp3 vol -0.4 amplitude

sox s1.mp3 s2.mp3 DeuxSons1.mp3
sox s2.mp3 -v 0.6 s1_faible1.mp3 DeuxSons2.mp3

sox s1.mp3 s1_avec_silence1.mp3 pad 1
sox s1.mp3 s1_avec_silence2.mp3 pad 1 0.5
sox s1.mp3 s1_avec_silence3.mp3 pad 1 5000s@3 0.5

sox -m s3.mp3 s4.mp3 SommeSons.mp3

sox s1.mp3 s1_avec_offset.mp3 dcshift 0.05

sox s1.mp3 s1_a_l_envers.mp3 reverse

sox s3.mp3 s3_morceau.mp3 trim 1.5 2

sox s3.mp3 s3_faded.mp3 fade t 10 1:00 20

rem pause
```

– The first two commands apply gains to the signal contained in file `s1.mp3`; 6 dB in the first case (corresponding to a factor of 0.5) and 0.4 in the second case (corresponding to a reduction in amplitude of 60% and a phase shift of 180°).

– The next two commands place two signals in sequence. In the second case, the amplitude of the first signal is reduced by a factor of 0.6.

– The following three commands add null values (i.e. silence) to the signal contained in the file `s1.mp3`. The first command adds 1 s of silence at the beginning, the second adds 1 s at the beginning and 0.5 s at the end and the third additionally inserts 5,000 samples from the third second of the signal.

– The following command mixes the two signals `s3.mp3` and `s4.mp3`.

– The next command adds a DC component equal to 0.05.

– The following command reverses a signal (the end becomes the beginning, and vice versa).

– The next command extracts a section of the signal contained in the file `s3.mp3`, beginning at 1.5 s with a duration of 2 s. This command is used to easily extract sections of signals for analysis or modification.

– The final command is used to modify the amplitude of a signal. In this way, it is possible to vary the amplitude of a signal progressively, at the beginning and at the end. Parameter `-t` corresponds to a linear modulation. The second parameter indicates that the amplitude of the signal moves from 0 to its maximum value in the course of the first 10 s; the next parameter indicates that a decrease begins after 1 min, and the final parameter indicates that the linear decrease continues for more than 20 s. These last two parameters are optional. Other types of modulation are possible, as shown in the software user manual.

EXERCISE 1.2.– Write a command file that generates a file we will call `succession.mp3`, containing the 10 signals generated in Exercise 1.1, placed in sequence. Next, write another command file to generate a second file, `Sommesons.mp3`, containing the sum of the 10 files generated in Exercise 1.1. What is the analytical expression of this second signal? What mathematical property does it satisfy?

EXERCISE 1.3.– Write a command file to generate a recording of 1 min of signal equal to the sum of two sinusoids with frequencies of 440 and 441 Hz and with the same amplitude. What does this signal look like? Justify its appearance using one of the following two relationships:

$$\cos(\theta_1) + \cos(\theta_2) = 2\cos\left(\frac{\theta_1+\theta_2}{2}\right)\cos\left(\frac{\theta_1-\theta_2}{2}\right)$$

$$\sin(\theta_1) + \sin(\theta_2) = 2\sin\left(\frac{\theta_1+\theta_2}{2}\right)\cos\left(\frac{\theta_1-\theta_2}{2}\right)$$

Finally, write a command file to extract 3 s from the file

```
AriaCantilenaVillaLobosPetibon.mp3
```

beginning at 5 min 24 s, then listen to and visualize this extract using `Audacity`. How is this connected to the previous question?

EXERCISE 1.4.– First generate a file `Sinus.mp3` containing a sinusoid of frequency 440 Hz, then a file `Noise.mp3` containing random noise, both with a length of 7 s. Then, use a command such as

```
sox -m Sinus.mp3 -v 0.01 Noise.mp3 SinusNoised.mp3
```

to add the sinusoid to the noise, with the noise being attenuated by a factor of 0.01. Beyond what value of the attenuation factor is the noise no longer perceptible to the human ear?

Finally, use a command such as

```
sox -m Noise.mp3 -v 0.01 Sinus.mp3 SinusNoised.mp3
```

to add the noise to the sinusoid, with the sinusoid being attenuated by a factor of 0.01. Beyond what value of the attenuation factor is the sinusoid no longer perceptible to the human ear?

EXERCISE 1.5.– The aim of this exercise is to show that it is possible to generate relatively complex files, which may then be used for tests or monitoring.

Write a command file to generate an `mp3` file containing a recording of 2 s of the sum of six sinusoids of frequency f_0, $2\,f_0$, $3\,f_0$, $4\,f_0$, $5\,f_0$ and $6\,f_0$, where $f_0 = 523\,\text{Hz}$. Next, generate two other files obtained using values of 659 and 830 Hz, respectively, for f_0. Finally, generate a file containing the sequence of these three sounds, and then a last file that repeats the previous file three times[1].

For this exercise, a spreadsheet may be used with file `GenerationCode.xls`, which automatically generates a set of SOX instructions; these instructions may then be copied into a command file.

EXERCISE 1.6.– Write a command file that:

1) generates two sinusoidal signals with a duration of 10 s and frequencies of 12,000 and 12,440 Hz. These two signals should be stored in files with extension `.wav`.

2) calculates the sum of these signals and stores them in a third file with extension `.wav`.

1 In 1968, Jean-Claude Risset used this principle to create a perpetually ascending sound, which gives the impression of constantly becoming higher and higher.

Listen to the obtained signal (preferably using headphones). What do you hear? Justify this result[2].

1.4. Using other file formats

In the previous section, signals were stored in mp3 format files. However, SoX is able to handle a wide variety of file formats, both in terms of reading and writing. These formats, described in detail in the user manual, include .wav, used by the Windows operating system, and .dat, which corresponds to text files (coded in ASCII), in which the first line corresponds to the sampling frequency and the following lines contain two numbers for time and the value of the signal. The command file below (Generation3.bat)

```
rem generation d'un fichier texte au format .dat
rem F. Auger, IUT Saint-Nazaire, dep. MP, jan. 2010

sox s1.mp3 s1.dat silence 0 trim 0.1 10s
```

generates a file s1.dat containing 10 successive values (10s) of file s1.mp3 taken 0.1 s after the start of the signal (having removed the null values from the beginning of the file). The contents of file s1.dat are as follows:

```
; Sample Rate 48000
; Channels 1
               0       -0.68915808
  2.0833333e-005       -0.65037082
  4.1666667e-005       -0.60942747
       6.25e-005       -0.56646369
  8.3333333e-005       -0.52161656
   0.00010416667       -0.47505816
        0.000125       -0.42689275
   0.00014583333       -0.37733861
```

2 This technique is used by directional speakers to generate acoustic waves using ultrasonic transducers. See [MUI 11].

```
0.00016666667       -0.32652545
    0.0001875       -0.27462651
```

By creating files of this type, we are able to manipulate measurement signals of non-acoustic origin using SoX. We must simply normalize these signals so that they always fall between -1 and 1. By choosing a sampling frequency suited to human audition, we are able to listen to non-acoustic phenomena. The program below (called `GenerationFichierDat2010.ch`), written in C for the Ch environment, reads numbers from a text file, extracts the maximum and the minimum, and then generates a `.dat` file, applying cross-multiplication to the values of the text file to convert values between x_{min} and x_{max} into values between -1 and 1. The result may then be converted into an `mp3` file, and may be listened to using the command

```
sox -V4 Mesures.dat Mesures.mp3
```

```
//
Programme permettant de transformer un fichier de mesures au format texte
//  en fichier de mesures au format .dat
//  F. Auger, janvier 2010

#include <stdio.h>

int main()
{
 char   NomFichierTexte[] = "Mesures.txt" ;
 char   NomFichierDat[] = "Mesures.dat" ;
 FILE   *PointeurFichierTexte, *PointeurFichierDat ;
 double x, xmax, xmin, xnorm, Somme, Difference, Fe=16000.0, Te=1.0/Fe, t;

 // on ouvre une premiere fois le fichier pour trouver le min et le max
 PointeurFichierTexte = fopen(NomFichierTexte, "r") ;
 if ( PointeurFichierTexte == NULL )
  {printf("Impossible d'ouvrir le fichier %s en lecture \n",
          NomFichierTexte);}
 else
  {
   // premiere lecture dans le fichier pour initialiser xmin et xmax
   if (fscanf(PointeurFichierTexte,"%lf", &x) != EOF)
    {xmin=x ; xmax=x;}
```

```
  while (fscanf(PointeurFichierTexte,"%lf", &x) != EOF)
   {
    if (x>xmax) {xmax=x;}
    if (x<xmin) {xmin=x;}
   }
  fclose(PointeurFichierTexte);

  Somme=xmax+xmin; Difference = xmax-xmin; t=0;

  // generation du fichier .dat par une relecture du fichier texte
  PointeurFichierTexte=fopen(NomFichierTexte,"r");// ouverture en lecture
  PointeurFichierDat =fopen(NomFichierDat,
"w");// ouverture en ecriture

  if ( PointeurFichierDat == NULL )
   {printf("Impossible d'ouvrir le fichier %s en ecriture\n",
          NomFichierDat);}
  else
   {
    // on precise la frequence d'echantillonnage du signal de sortie, qui
    // n'est pas forcement la meme que celle du signal d'entree
    fprintf(PointeurFichierDat,"; Sample Rate %8.2f\n", Fe);

    // un seul signal = une seule voie = un seul canal
    fprintf(PointeurFichierDat,"; Channels 1\n");

    while (fscanf(PointeurFichierTexte,"%lf", &x) != EOF)
     {
      xnorm=(2*x-Somme)/Difference;
      fprintf(PointeurFichierDat, "%f %f \n", t, xnorm);
      t=t+Te;
     }

    fclose(PointeurFichierTexte);// c'est fini, on ferme les fichiers
    fclose(PointeurFichierDat);
   }
  }
 return 0;
}
```

EXERCISE 1.7.– If x is between $x_{\min}$ and $x_{\max}$, which intervals must contain

$$\frac{x - x_{\min}}{x_{\max} - x_{\min}} \quad \text{and} \quad 2\,\frac{x - x_{\min}}{x_{\max} - x_{\min}} - 1 \quad ?$$

Provide a justification of the expression of `xnorm` used in the program. Next, modify the previous program to convert the file `ecg.txt` into `.dat` format. This file contains an electrocardiogram sampled at 720 Hz and acquired using a 12 bit analog–digital converter[3]. This file may then be converted into `.wav` format using SoX, then visualized using `Audacity`.

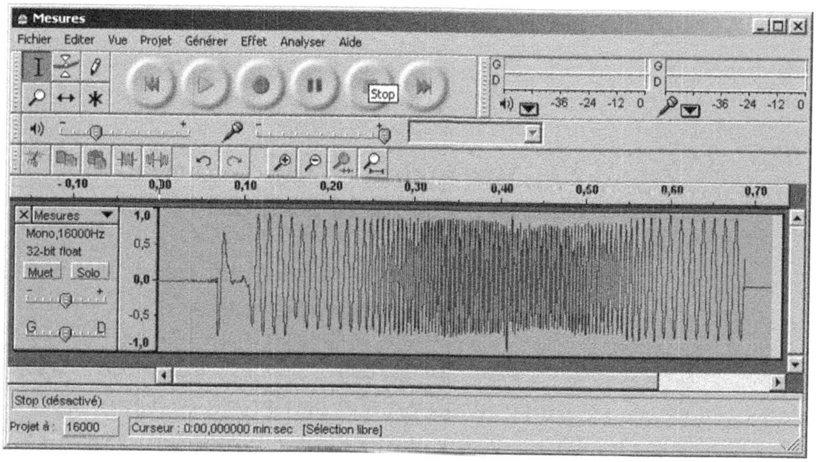

Figure 1.1. *Result of the creation of an* `mp3` *file based on electrical currents measured at the terminals of an asynchronous machine*

This format may also be used to process signals generated by SoX in other programs, for example in a spreadsheet program such as `LibreOffice Calc`[4].

3 Source: http://www.physionet.org/physiobank/database/aami-ec13/.
4 See http://fr.libreoffice.org/.

2

Signal Analysis

2.1. Measuring the temporal and frequency characteristics of a signal

Three commands may be used to analyze a signal using certain specific characteristics:

– The program soxi, supplied with SoX, provides information on the structure of a file. Thus, the command file (Analyse1.bat)

```
rem analyse d'un fichier avec soxi
rem F. Auger, IUT Saint-Nazaire, dep. MP, jan. 2010

soxi son1.mp3
soxi son1.mp3 > son1_info_soxi.txt
pause
```

provides the information shown below. With the first command, this information is shown on-screen. Using the second command, the information is stored in a file son1_info_soxi.txt.

```
Input File      : 'son1.mp3'
Channels        : 1
Sample Rate     : 48000
Precision       : 16-bit
Duration        : 00:00:03.55 = 170496 samples ~ 266.4 CDDA
```

```
                  sectors
File Size       : 28.4k
Bit Rate        : 64.0k
Sample Encoding: MPEG audio (layer I, II or III)
```

– The `stat` command in SoX determines certain characteristic parameters of a signal. Thus, the command file[1] (`Analyse2.bat`)

```
rem analyse d'un fichier avec sox stat
rem F. Auger, IUT Saint-Nazaire, dep. MP, jan. 2010

sox son1.mp3 -n stat
sox son1.mp3 -n stat 2> son1_info_stat.txt
pause
```

produces the information shown below. With the first command, this information is shown on-screen. Using the second command, the information is stored in a file `son1_info_stat.txt`.

```
Samples read:               169344
Length (seconds):         3.528000
Scaled by:            2147483647.0
Maximum amplitude:        0.983058
Minimum amplitude:       -0.977098
Midline amplitude:        0.002980
Mean    norm:             0.599965
Mean    amplitude:        0.000013
RMS     amplitude:        0.669044
Maximum delta:            0.059710
Minimum delta:            0.000000
Mean    delta:            0.034553
RMS     delta:            0.038529
Rough   frequency:             439
```

1 Warning: no space between “2” and “>”. Specialists may wish to note that the normal output from SoX is `stderr` and not `stdout`. The syntax `2>` allows us to redirect the `stderr` flow to a file, while the command `>` or `1>` redirects the `stdout` flow. In the absence of redirection, the results appear on the screen.

```
Volume adjustment:          1.017

Try: -t raw -U -1
```

– Finally, the command `stats` in SoX provides other characteristic parameters of a signal. Thus, the command file (`Analyse3.bat`)

```
rem analyse d'un fichier avec sox stats
rem F. Auger, IUT Saint-Nazaire, dep. MP, jan. 2010

sox son1.mp3 -n stats
sox son1.mp3 -n stats 2> son1_info_stats.txt
pause
```

produces the information shown below. With the first command, this information is shown on-screen. Using the second command, the information is stored in a file `son1_info_stats.txt`.

```
DC offset     0.000013
Min level    -0.977098
Max level     0.983058
Pk lev dB        -0.15
RMS lev dB       -3.49
RMS Pk dB        -3.44
RMS Tr dB        -3.89
Crest factor      1.47
Flat factor       0.00
Pk count             2
Bit-depth        29/29
Num samples       169k
Length s         3.528
Scale max     1.000000
Window s         0.050
```

The SoX user manual provides information on the nature of these characteristic parameters, which are too numerous to discuss in detail here; some are also redundant.

EXERCISE 2.1.– Analyze and verify the results obtained using command file[2] below (`Analyse20130624.bat`), paying particular attention to the effective value of the signal.

```
rem analyse de signaux elementaires avec SoX
rem F. Auger, IUT Saint-Nazaire, dep. MP, june 2013

sox -n ana_s1.wav synth 5 sine 440
sox -n ana_s2.wav synth 5 triangle 440
sox -n ana_s3.wav synth 4 square 440 0 0 40

sox ana_s1.wav -n stat 2>  AnaStat.txt
sox ana_s2.wav -n stat 2>> AnaStat.txt
sox ana_s3.wav -n stat 2>> AnaStat.txt

rem pause
```

EXERCISE 2.2.– What is the sampling frequency of the signals generated in Exercise 1.1? Write a command file to produce a highly detailed analysis of file `SonRigolo3.mp3`. Discuss the obtained results.

2.2. Time-frequency analysis of a signal

The structure of certain signals may be clearly highlighted using a time-frequency representation. This representation corresponds to a distribution of the signal energy in a plane made up of the temporal and frequency dimensions. In this way, we are able to visualize the way in which the frequency representation of the signal evolves over time. The obtained image may be compared to a music score, with time shown on the x axis and frequency on the y axis. For example, the command file (`Analyse4.bat`)

---

2 The use of the command $\gg$ allows us to place all results into the same file: the output flow is redirected to the end of the file `AnaStat.txt`.

```
rem analyse d'un fichier avec sox spectrogram
rem F. Auger, IUT Saint-Nazaire, dep. MP, jan. 2010

sox -n s6a.wav synth 3 sine  660-2640
sox -n s6b.wav synth 3 sine 1320-5280
sox -n s6c.wav synth 3 sine 1980-7920
sox -m s6a.wav s6b.wav s6c.wav s6.wav

sox s6.wav -n spectrogram        -o s6_sp.png
sox s6.wav -n spectrogram -m     -o s6_sp2.png
sox s6.wav -n spectrogram -l     -o s6_sp3.png
sox s6.wav -n spectrogram -l -m -o s6_sp4.png
sox s6.wav -n spectrogram -l -m -S 0.5 -d 1.3 -o s6_sp5.png

rem pause
```

generates a file `s6.mp3` containing a recording of 3 s of the sum of three sinusoids of which the frequencies increase over time. This signal is then analyzed using a spectrogram, which is a specific method of time-frequency analysis. The results of this analysis are stored in the output file (`-o`) `s6_sp.png`. The following command generates the same spectrogram in grayscale rather than in color, a format suitable for printing using a black-and-white printer. The next two commands use option `-l`,[3] which produces a white background. The final command analyzes the signal starting (`-S`) from instant 0.5 s over a duration (`-d`) of 1.3 s. The results of the penultimate command are presented in Figure 2.1. Among other things, this representation clearly shows the fundamental component and its harmonics, with frequencies that increase progressively over time. The SoX user manual gives a detailed description of the various parameters of the `spectrogram` command.

3 Note that "l" is the lowercase letter L and not the number 1 (one).

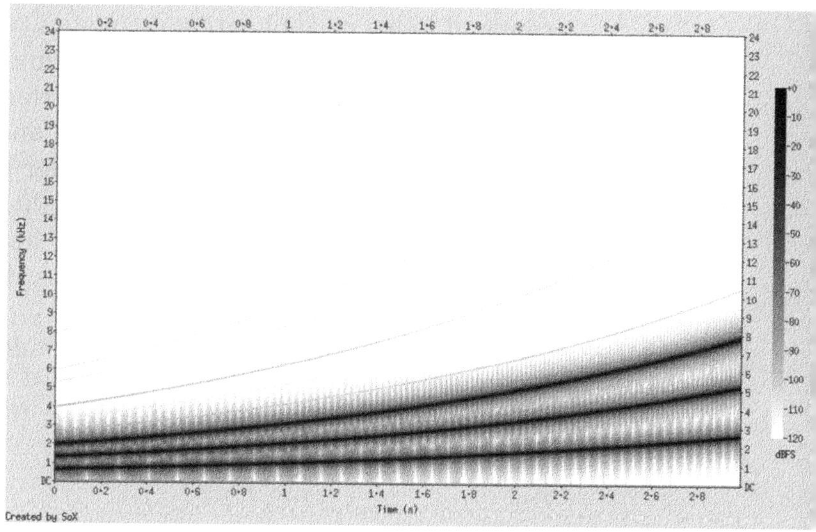

Figure 2.1. *Time-frequency representation of signal* `s6.wav`, *obtained using a spectrogram.* `dBFS` *stands for dB full scale and indicates that the different shades of gray correspond to the value in dB of the spectrogram, normalized by its maximum value*

EXERCISE 2.3.– Use spectrograms to differentiate the three signals generated by the following commands:

```
sox -n chirp1.wav synth 3 sine  1000:20000
sox -n chirp2.wav synth 3 sine  1000+20000
sox -n chirp3.wav synth 3 sine  1000/20000
```

EXERCISE 2.4.– Complete the command file from Exercise 2.2 to construct a spectrogram of signal `SonRigolo3.mp3`, both as a whole and for specific parts of the signal. Discuss your results.

2.3. Non-parametric spectral analysis

Spectral analysis, the representation of the distribution of energy of a signal in the frequency domain, is used in a wide range of fields of application (acoustics, vibration mechanics, biomedical engineering, telecommunications, astronomy, earth sciences, etc.). Based on the Fourier transform, spectral analysis is an efficient tool for the analysis of stationary signals. In this section, we will consider its use for a number of signals to demonstrate the possibilities it offers, clearly highlighting the fact that the field of application of spectral analysis is far wider than the examples shown below.

A possible starting point for the presentation of this technique is the Fourier transform of a signal $x(t)$, which is the function of the real variable f (the frequency) with complex values defined by:

$$X(f) = \int_{-\infty}^{+\infty} x(t)\, e^{-\jmath 2\pi f t}\, dt$$

As the signal is only observed and recorded over a finite length of time L from an initial instant t_0, we wish to approximate $X(f)$ using the Fourier transform of the observed part of $x(t)$,

$$\begin{aligned} \hat{X}(f) &= \int_{t_0}^{t_0+L} h(t - t_0)\, x(t)\, e^{-\jmath 2\pi f t}\, dt \\ &= \int_0^L h(u)\, x(t_0 + u)\, e^{-\jmath 2\pi f u}\, du \end{aligned}$$

where $h(t)$ is a weighting function chosen by the user and used to specify the importance of $x(t)$ in the calculation of $\hat{X}(f)$. This integral is then calculated in an approximate manner by subdividing the interval $[0; L]$ into N subintervals of the same width $T_s = L/N$ and approximating each integral using the rectangle method:

$$\begin{aligned}\hat{X}(f) &= \sum_{n=0}^{N-1} \int_{n\,T_s}^{(n+1)\,T_s} h(u)\,x(t_0+u)\,e^{-\jmath 2\pi f u}\,du \\ &\approx Te \sum_{n=0}^{N-1} h(n\,T_s)\,x(t_0+n\,T_s)\,e^{-\jmath 2\pi f n\,T_s} \\ &= \frac{L}{N} \sum_{n=0}^{N-1} h(n\,T_s)\,x(t_0+n\,T_s)\,e^{-\jmath 2\pi f n\,T_s}\end{aligned}$$

The most widespread practice is to calculate this frequency representation at frequencies $f = m\,\frac{F_e}{N}$. This gives us the expression

$$\tilde{X}\left(m\,\frac{F_s}{N}\right) = \frac{L}{N} \sum_{n=0}^{N-1} h(n\,T_s)\,x(t_0+n\,T_s)\,e^{-\jmath 2\pi \frac{mn}{N}}$$

which is often calculated using an algorithm known as the fast Fourier transform (FFT), for which we graphically represent the square module $|\tilde{X}(f)|^2$ (or its value in dB) as a function of the frequency. The number of samples N thus determines the duration of the recording, $L = N\,T_s$, and the discretization step of the frequency axis, $\Delta f = F_s/N$.

2.3.1. *Analytical case studies*

EXERCISE 2.5.– We wish to calculate the Fourier transform of a recording of finite length of a sinusoidal signal with frequency f_0, defined by $x(t) = A\cos(2\pi f_0\,t+\varphi)$. We presume that the recording of this signal is of length L, begins at $t_0 = -L/2$ and ends at $+L/2$. A rectangular window will be selected as the "smoothing window" for the signal:

$$h(t) = \begin{cases} 1 & \text{if} \quad -L/2 \leq t \leq +L/2 \\ 0 & \text{else} \end{cases},$$

Show that the Fourier transform of this sinusoid segment may be written in the form:

$$\hat{X}(f) = \tfrac{1}{2}\, A\, e^{+\jmath\varphi}\, H(f - f_0) + \tfrac{1}{2}\, A\, e^{-\jmath\varphi}\, H(f + f_0).$$

Specify the expression of $H(f)$ and calculate $H(0)$. If the amplitude A of the sinusoid is estimated by measuring $\frac{2|\hat{X}(f_0)|}{H(0)}$, give a majorant of the absolute value of the relative error. This may be obtained using the following triangular inequality:

$$|x| - |y| \leq |x + y| \leq |x| + |y|$$

Discuss the practical applications of the obtained result.

EXERCISE 2.6.– Let the signal $x(t)$ be equal to zero at negative times and to $e^{-t/T}$ at positive times:

$$x(t) = e^{-t/T}\Gamma(t) = \begin{cases} e^{-t/T} & \text{if} \quad t \geq 0 \\ 0 & \text{else} \end{cases}$$

We wish to perform a spectral analysis on a recording of that signal, whose length is L. This entails calculating the Fourier transform of a signal $y(t)$ equal to $x(t)$ if $0 \leq t \leq L$, and equal to zero otherwise. Calculate the Fourier transforms of $x(t)$ and $y(t)$. Write the relation which must be satisfied by the length L of the recording in order for the maximum relative error committed by confusing $Y(f)$ and $X(f)$ to be less than 1%.

EXERCISE 2.7.– Calculate the Fourier transform of the signal

$$x(t) = \begin{cases} A\frac{t}{T}e^{-t/T} & \text{for } t > 0 \\ 0 & \text{for } t < 0 \end{cases},$$

where $A \in I\!R_+$ and $T \in I\!R_+^*$. From this, deduce the Fourier transform of the signal

$$y(t) = \begin{cases} A\frac{t}{T}e^{-t/T} & \cos(2\pi f_0 t) & \text{for } t > 0 \\ 0 & & \text{for } t < 0 \end{cases},$$

where $A \in I\!R_+$, $f_0 \in I\!R_+$ and $T \in I\!R_+^*$. An example of this type of signal is shown in Figure 2.2.

EXERCISE 2.8.–

1) Recall the definition of the Fourier transform $X(f)$ of a signal $x(t)$. Show that $\frac{dX}{df}(f)$ and $\frac{d^2X}{df^2}(f)$ are, to within certain multiplication factors which we shall determine, the Fourier transforms of $t\, x(t)$ and $t^2\, x(t)$.

2) The Fourier transform of $x(t) = e^{-\frac{t^2}{2}}$ is $X(f) = \sqrt{2\pi}e^{-2\pi^2 f^2}$. By changing a variable, deduce from this the expression of the Fourier transform of $y(t) = e^{-\frac{t^2}{2T^2}}$, with T being a real, positive parameter characteristic of $y(t)$.

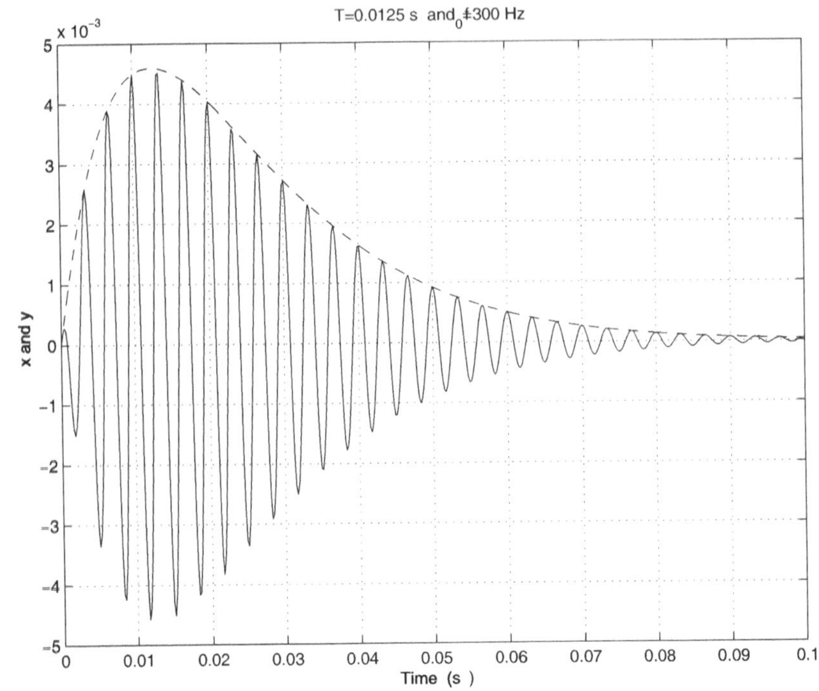

Figure 2.2. *Signals $x(t)$ (dotted lines) and $y(t)$ (solid line) mentioned in Exercise 2.7*

3) With reference to the previous two questions, calculate the Fourier transform of the Ricker wavelet (also known as the "Mexican hat wavelet" – see Figure 2.3), defined by

$$h(t) = \frac{1}{\sqrt{2\pi}T}\left(1 - \frac{t^2}{T^2}\right)e^{-\frac{t^2}{2T^2}}$$

2.3.2. *Practical application*

EXERCISE 2.9.– Using the command file below (which corresponds to `AnalyseSpectrale1.bat`), we may generate a signal which is equal to the sum of two sinusoids with frequencies of 400 and 500 Hz. For how many samples must the spectral analysis of this signal be calculated in order to distinguish between the lines of the two frequencies in the spectrum produced by `Visual Analyser`? Examine the effects of the choice of a weighting window and of the number of averages on the obtained representation.

```
rem generation de signaux elementaires avec SoX
rem F. Auger, IUT Saint-Nazaire, dep. MP, fev. 2010

sox -n AnaSpec1.mp3 synth 3:00 sine 400
sox -n AnaSpec2.mp3 synth 3:00 sine 500

sox -m AnaSpec1.mp3 AnaSpec2.mp3 AnaSpec4.mp3

rem pause
```

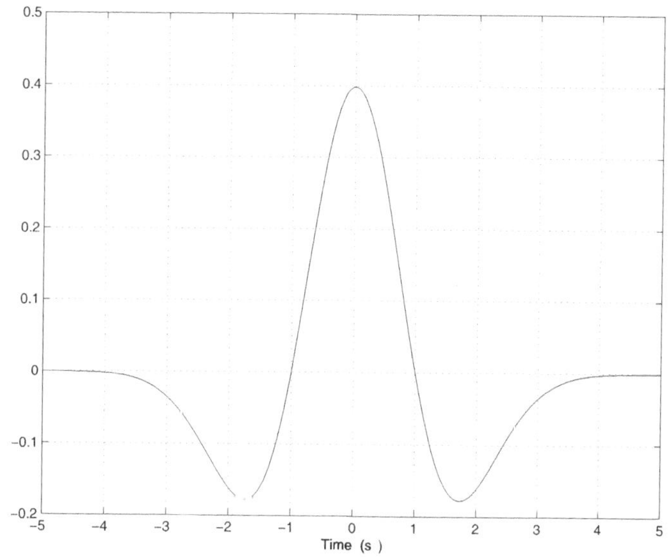

Figure 2.3. *Graphic representation of the Ricker wavelet (with $T = 1$ s) studied in Exercise 2.8*

EXERCISE 2.10.– Using SoX and `Visual Analyser`, analyze signal `SignalStationnaire1.mp3` and a second signal, `SignalStationnaire2.mp3`, in order to identify the content as precisely as possible. Once you have specified the parameters of your spectral analysis, do not hesitate to use recording functions (`settings/capture/capture spectrum`), and particularly spectrum recording in a text file (`File/Save Spectrum as text`).

3

Signal Processing

3.1. Sampling

A clear understanding of effective signal sampling conditions is essential for correct manipulation and processing of discrete-time signals. Sampling is an operation that converts a continuous-time signal $x(t)$ into a time-discrete signal $x[n]$, obtained by taking the value of $x(t)$ at multiple instants over the sampling period: $x[n] = x(n T_/Delta)$. It is therefore easy to pass from $x(t)$ to $x[n]$. We will now consider the opposite situation, i.e. the reconstruction of $x(t)$ based on $x[n]$. Using simple sinusoid signals, we can clearly demonstrate that sampling leads to a loss of information:

– The signal $x(t) = A \cos(2\pi(f + F_s)t + \varphi)$, where $F_s = 1/T_s$ is known as the sampling rate, produces the following signal after sampling:

$$
\begin{aligned}
x[n] &= A \cos(2\pi f T_s n + 2\pi F_s T_s n + \varphi) \\
&= A \cos(2\pi f T_s n + \varphi)
\end{aligned}
$$

After sampling, this sinusoid of frequency $f + F_s$ will be considered as another sinusoid of frequency f. This applies to all sinusoids of frequency $f + k F_s$, where k is a positive integer. This phenomenon, which is due to the periodicity of trigonometric functions, is known as *frequency aliasing by translation*. This phenomenon can cause the sampling process to modify the frequency of a sinusoid. To avoid this possibility, the frequency must be lower than F_s.

– The two sinusoid signals $z_1(t) = A\cos\left(2\pi\left(\frac{F_s}{2}+\Delta f\right)t\right)$ and $z_2(t) = A\cos\left(2\pi\left(\frac{F_s}{2}-\Delta f\right)t\right)$, corresponding to two sinusoids of the same amplitude and with symmetrical frequencies with respect to $F_s/2$, produce the same discrete-time signal $z_1[n] = z_2[n] = (-1)^n A\cos(2\pi\Delta f T_s n)$ after sampling. This phenomenon is a result of the elementary properties of trigonometric functions (particularly the expressions of $\cos(\theta_1+\theta_2)$ and $\cos(\theta_1-\theta_2)$), and is known as *frequency aliasing by symmetry*. To prevent modification of the frequency of a sinusoid as a result of this sampling process, the frequency must be lower than $F_s/2$.

On the basis of these two simple results, C. E. Shannon demonstrated that any given signal $x(t)$ may be sampled correctly at sampling rate F_s if the following two conditions are satisfied:

– A frequency $f_{\max}$ exists such that $X(f)$, the Fourier transform of $x(t)$, is null for any frequency f greater than $f_{\max}$. This means that the signal does not include energy located above $f_{\max}$, which is the highest frequency contained in the signal.

– The selected sampling rate F_s must be at least twice $f_{\max}$: $F_s \geq 2f_{\max}$.

If these two conditions are verified, it is possible to calculate the value of the continuous-time signal $x(t)$ at any instant t using the values of $x[n] = x(nT_s)$, using an expression known as *Shannon's interpolation formula*.

In conclusion, for a signal to be correctly sampled at frequency F_s, two conditions must be satisfied:

– All of the *information* contained in the signal must be below $F_s/2$. If this condition is not satisfied, we must either change the sampling rate or accept a loss of information.

– No significant *energy* should be located above $F_s/2$. If this condition is not satisfied, we must use a low-pass filter to verify Shannon's theorem. This is known as an anti-aliasing filter.

3.1.1. *Analytical case studies*

EXERCISE 3.1.– A microprocessor constantly repeats a cycle comprising a measurement of the voltage present at the input to an analog-to-digital converter (ADC) and a recording of the result in a digital-to-analog converter (DAC). Its operation rate is such that it can perform 2000 such cycles every second.

A sinusoidal voltage of frequency 1500 Hz is then applied at the input to the ADC, using a low frequency (LF) signal generator.

What is the frequency of the signal output by the DAC?

EXERCISE 3.2.– Figure 3.1(a) shows the samples of a sinusoidal signal whose frequency is $f_0 = 396$ Hz. The time-period between two successive samples is 10 ms. Is this signal correctly sampled? Analyze the result obtained if need be.

Figure 3.1(b) shows the samples of a sinusoidal signal whose frequency is $f_0 = 112$ Hz. This signal was recorded over 0.6 s, providing 601 points. Is this signal correctly sampled? Analyze the result obtained if need be.

Figure 3.1(c) shows the samples of a sinusoidal signal whose pulsation is $\omega_0 = 250$ rad/s, sampled at a rate of 90 points per second. Is this signal correctly sampled? Analyze the result obtained if need be.

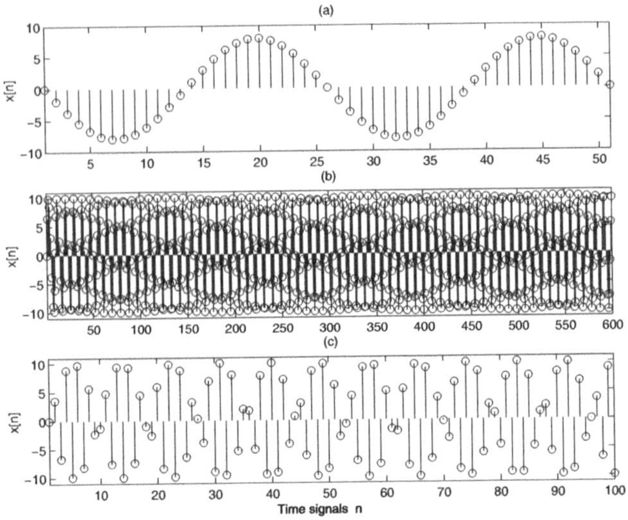

Figure 3.1. *Temporal representations of the discrete-time signals studied in Exercise 3.2*

EXERCISE 3.3.– By decomposing it into a Fourier series, it is possible to consider any periodical signal with period T_0 as the sum of a constant term (the continuous component) and sinusoidal components, of multiple frequencies with the base frequency $f_0 = \frac{1}{T_0}$. If we sample at 300 Hz (without an anti-aliasing filter) a periodic signal whose base frequency is $f_0 = 45$ Hz, at what frequencies will its first 10 sinusoidal components be perceived after sampling? You can fill in the table shown in Figure 3.2, but justify your answers. What will the periodicity of this signal be after sampling?

EXERCISE 3.4.– We wish to acquire a signal in the form $x(t) = A \sin(\frac{2\pi t}{T})$ to measure its amplitude A.

1) At what time can this parameter be measured directly?

2) The signal is sampled with a sampling period T_s, and we try to deduce A from the maximum value of all the $x[n]$ values. Figure 3.3 shows the sampled signals obtained for different values of T_s. In what circumstances will the error in the measurement of A be greatest, and why?

3) Consider the case of the first column in Figure 3.3, i.e. where the two values closest to A are obtained at times $\frac{T}{4} - \frac{T_s}{2}$ and $\frac{T}{4} + \frac{T_s}{2}$. We wish to find the value of T_s for which the error committed in this case will be less than 0.5%. Deduce from this an inequality between T and T_s, and another linking $F_e = \frac{1}{T_s}$ and $f = \frac{1}{T}$. What do these inequalities become if we want the maximum error to be less than 0.1%? Draw a conclusion from the results obtained.

k	frequency before sampling	frequency after sampling	k	frequency before sampling	frequency after sampling
1	45		6		
2	90		7		
3			8		
4			9		
5			10	450	

Figure 3.2. *Table used in Exercise 3.3*

EXERCISE 3.5.– A sensor delivers a signal which contains pertinent data in a frequency interval $[0, B]$. This signal is first run through a low-pass anti-aliasing filter with cutoff frequency f_c, and then sampled at a frequency F_s. We choose $f_c = 1.5\,B$ and $F_s = 3\alpha\,B = 2\alpha\,f_c$, where $\alpha > 1$. The objective of this exercise is to determine the value of α for which the components of the signal situated in the interval $[0, B]$ are least altered.

1) Let f be a frequency in the interval $[0, B]$. What is the lowest frequency f' which will modify the frequential response of the signal with frequency f by superposition at that frequency by spectral aliasing?

2) The anti-aliasing filter used is an N-order Butterworth filter, the modulus of whose frequential response,

$$|H(f)| = \frac{1}{\sqrt{1 + (f/f_c)^{2N}}}$$

is a decreasing function on $\mathbb{R}_+$. What is the frequency f of the interval $[0, B]$ which would be disturbed by the frequency f' which is least attenuated by the filter?

3) At the frequency f' found in the previous question, we want the signal to be attenuated by a factor of 100. Deduce the value of α for N ranging from 1 to 5. Draw a conclusion about the results obtained.

4) The `Tektronix 2642A` analyzer uses a 6th-order Butterworth filter with $F_s = 2.56\, f_c$. What is the attenuation at the frequency f'?

EXERCISE 3.6.– For a digital oscilloscope, the sample rate is an important parameter.[1] In the technical specifications of the devices, it is always the maximum sample rate that is specified. However, in certain cases, the true sample rate may be considerably lower, for two reasons:

– the number of acquisition channels. Often, digital oscilloscopes have only one ADC, which is therefore preceded by an analog multiplexer to take acquisitions on multiple channels. In twin-trace mode (two channels), the sample rate is therefore halved;

1 See the article “Oscilloscopes: comment choisir la fréquence d'échantillonnage?”, *Mesures*, No 760, pp 38–40, December 2003.

– the size (or "depth") of the memory, i.e. the number of samples that the oscilloscope can record during an acquisition. The oscilloscope thus calculates the sample rate so that the whole of its memory is used for an acquisition.

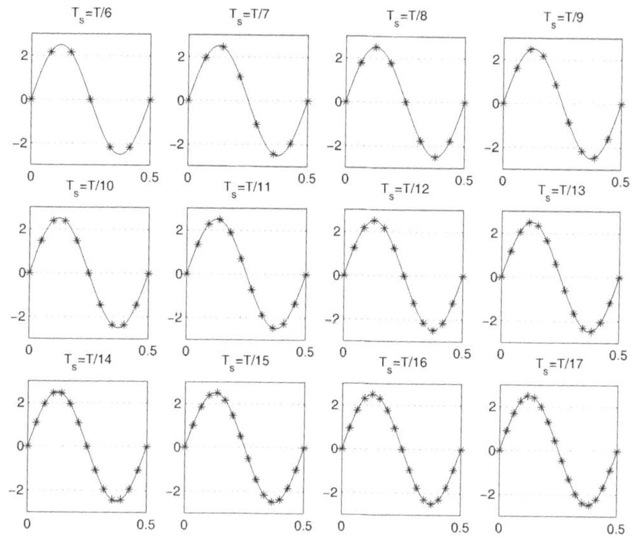

Figure 3.3. *Sampled signals $x[n]$ (*) obtained from the signal $x(t)$ (solid line) in Exercise 3.4 for T_s ranging from $T/6$ to $T/17$, with $T = 0.5$ s and $A = 2.5$*

Hence, for example, with a digital oscilloscope with a memory depth of 2 Mega Samples ($2 \times 1024 \times 1024 = 2097152$ samples),

– the recording of one signal with a timebase of 100 μ s/div will take $10 \times 100\ \mu\text{s} = 1$ ms, because there are 10 horizontal graduations on the screen of the oscilloscope. The sample rate, i.e. the number of samples per second, will therefore be

$$
\begin{aligned}
F_e &= \frac{\text{memory depth}}{(\text{timebase}) \times (\text{number of graduations})} \\
&= \frac{2 \times 1024 \times 1024}{10^{-3}} = 2.097 \text{ GHz}
\end{aligned}
$$

– the recording of two signals (twin-trace mode) with the same timebase will take the same length of time, but the need to store two signals in the memory instead of one, and the presence of the multiplexer before the ADC, mean that the sample rate is halved, so $F_s = 1.048$ GHz.

We want to use a digital oscilloscope, with a 24-Mega Samples memory depth and a maximum sample rate of 20 GHz to visualize signals containing transitory phenomena, oscillating at 700 MHz.

1) Is it possible to completely and correctly visualize this kind of signal on a single channel using a timebase of 100 μs/div?

2) Is it possible to completely and correctly visualize these transitory phenomena if we take an acquisition on one channel with a timebase of 1 ms/div?

3) Is it still possible if we take an acquisition on two channels with the same timebase? How will the transitory phenomena then appear?

EXERCISE 3.7.– We wish to sample a real signal whose energy is located solely in a frequency band of bandwidth B, centered on the frequency f_c, so between $f_c - B/2$ and $f_c + B/2$.

1) Given the normal rules of sampling, how must we choose the sample rate F_s so that this signal is correctly sampled? For $f_c = 10$ kHz and $B = 80$ Hz, why is this solution undesirable?

2) By application of the principle of frequency translation, what sample rate F_s must we choose so that the component of the signal with frequency $f_c - B/2$ appears in the same way as a constant signal (zero frequency)?

3) If the sample rate used satisfies the relation obtained in the previous question, what additional relation must F_s

and B satisfy so that there is no frequency overlap due to spectral aliasing? Why is this the optimum solution? Find a value of F_s which is appropriate for the values of f_c and B given previously.

3.1.2. *Practical applications*

EXERCISE 3.8.– The program below (which corresponds to file `Echantillonnage.ch`), written in C for the scientific computing environment `Ch`, generates a file in `.dat` format containing a 20 s recording of a sinusoid with a frequency that varies from $f_i = 400$ Hz to $f_f = 6,000$ Hz.

```
// Programme permettant de generer la version echantillonnee
// d'une sinusoide de frequence croissante
// F. Auger, janvier 2010

#include <stdio.h>
#include <math.h>

int main()
{
 char   NomFichierDat[] = "Echantillonnage.dat" ;
 FILE   *PointeurFichierDat ;
 double MySignal, Fe, Te, t=0.0, Duree, fi, ff, Omegai, alpha ;
 double Amp=0.9, Phi, MyPi ;

 Duree=20.0;              // duree du signal en secondes
 Te=40e-6 ; Fe=1.0/Te;    // periode et frequence d'echantillonnage
 fi=400.0, ff=6000.0;     // frequences initiales et finales

 MyPi=M_PI; Omegai = 2*MyPi*fi; alpha=2*MyPi*(ff-fi)/Duree;

 // ouverture du fichier en ecriture
 PointeurFichierDat = fopen(NomFichierDat, "w") ;

 if ( PointeurFichierDat == NULL )
  {printf("Impossible d'ouvrir le fichier %s en ecriture\n",
          NomFichierDat);
  }
 else
  {
```

```
    // on precise la frequence d'echantillonnage du signal de sortie
    fprintf(PointeurFichierDat,"; Sample Rate %8.2f\n", Fe);

    // un seul signal = une seule voie = un seul canal
    fprintf(PointeurFichierDat,"; Channels 1\n");

    while (t<Duree)
      {
       Phi=Omegai*t+0.5*alpha*t*t; // phase instantanee
       MySignal=Amp*cos(Phi);      // calcul du signal
       fprintf(PointeurFichierDat, "%f %f \n", t, MySignal);
       t=t+Te;
      }

     fclose(PointeurFichierDat);    // c'est fini, on ferme
   }

  return 0;
}
```

The sinusoid signal takes the form $x(t) = 0.9\cos(\varphi(t))$, with $\varphi(t) = \omega_i\, t + \alpha\, t^2/2$. Calculate $\Omega(t) = \frac{d\varphi}{dt}(t)$ and $\frac{\Omega(t)}{2\pi}$. How does the frequency vary as a function of time? Launch the program with a sampling period of $T_s = 40\ \mu$s and listen to the result. The `.dat` file may be transformed into a `.wav` file using the command

```
rem conversion d'un fichier .dat en .mp3
rem F. Auger, IUT Saint-Nazaire , dep. MP, jan. 2010

sox Echantillonnage.dat Echantillonnage.wav
```

Relaunch the program with a sampling period of $125\ \mu$s, and listen to the results again. Run a spectrogram analysis on the obtained signal. What has happened? How does this relate to Shannon's theorem?

EXERCISE 3.9.– Use the program from Exercise 3.8 to generate a sinusoid signal with a duration of 20 s, for which the frequency increases linearly from 400 to 8,000 Hz, sampled using a period of $170\ \mu$s. Convert the resulting file to

.wav format. Listen (using headphones) and produce a spectrogram of the results. Analyze and explain your results.

3.2. Resampling

SoX only handles sampled signals. All signals recorded using digital media are discrete-time signals. The program offers the possibility of resampling signals by changing the sampling rate; this constitutes one of the highlights of SoX. The operation is carried out using the rate command, as shown below:

```
sox SonRigolo.mp3 SonRigolo2.mp3 rate 32k
soxi SonRigolo.mp3  >  SonRigolo_info.txt
soxi SonRigolo2.mp3 >> SonRigolo_info.txt
```

This produces a new signal SonRigolo2.mp3 by resampling[2] the signal contained in file SonRigolo.mp3 at 32 kHz, instead of the original sampling rate of 44.1 kHz. The two subsequent soxi commands produce the file below, which clearly shows the sample rate of the two files:

```
Input File     : 'SonRigolo.mp3'
Channels       : 1
Sample Rate    : 44100
Precision      : 16-bit
Duration       : 00:00:03.58 = 157834 samples = 268.425 CDDA sectors
File Size      : 28.6k
Bit Rate       : 64.0k
Sample Encoding: MPEG audio (layer I, II or III)

Input File     : 'SonRigolo2.mp3'
Channels       : 1
Sample Rate    : 32000
Precision      : 16-bit
Duration       : 00:00:03.60 = 115200 samples ~ 270 CDDA sectors
File Size      : 21.6k
```

2 When using the MP3 standard, the authorized sample rates are 48 kHz, 44.1 kHz, 32 kHz, 24 kHz, 22,050 Hz, 16 kHz, 12 kHz, 11,025 Hz and 8 kHz.

```
Bit Rate        : 48.0k
Sample Encoding: MPEG audio (layer I, II or III)
```

EXERCISE 3.10.– What operations does SOX need to carry out in order to correctly resample a signal with an original sampling rate of 40 kHz using a new sampling rate of: (1) $F_{s2} =$ 10 kHz? (2) F_{s2}=16 kHz?

EXERCISE 3.11.–

– Analyze signal `EmissionFranceInter.mp3` to determine the number of samples it contains and the original sample rate.

– Extract the first 2 min from this signal using the `trim` command, discussed in section 1.3, and save this extract in a `.wav` file.

– Resample this new signal using smaller and smaller sampling rate values, and produce a table showing the number of samples and the size of the obtained file (in bytes) as a function of the new sampling rate. Analyze your results. At what sample rate does the signal become unintelligible?

3.3. "Analog" filtering

Stationary linear systems are useful for conditioning measurement signals, i.e. highlighting the information carried by the signals. Generally, these systems are as follows:

– *Dynamic*: their evolution is governed by a differential equation (for continuous-time systems) or by a recurrence equation (for discrete-time systems).

– *Mono-variable*: they have a single input value (known as *excitation*) and a single output value (the *response*).

– *Causal*: the value of the output at instant t depends solely on the input value at instants $t' \leq t$. This property translates the fact that effects cannot precede causes.

– *Linear*: they verify the superposition principle, which states that the response of a system to a sum of elementary excitations is the sum of the responses of the system to each of these excitations taken individually.

– *Stationary*: if the same excitation is given as input on two successive occasions, in strictly identical initial conditions, the two responses obtained will also be strictly identical. This property reflects the fact that the same causes produce the same effects, excluding the possibility of evolution, by wear and tear or aging, of the system.

If a continuous-time system satisfies all of these properties, we may show that the output $y(t)$ may be deduced from the input $x(t)$ by an operation known as a *convolution product*:

$$y(t) = \int_0^{+\infty} h(\tau)\, x(t-\tau)\, d\tau$$

where $h(t)$ is a characteristic function of the system known as the impulsional response. One important consequence of this result is that the response of a stationary linear system to a sinusoid is a sinusoid with the same frequency, amplified and phase shifted by terms that depend only on the system in question and the frequency of the sinusoid:

$$\begin{array}{lrcl} \text{if} & x(t) & = & X\, e^{\jmath 2\pi f t}, \\ \text{then} & y(t) & = & H_{\mathcal{F}}(f)\, X\, e^{\jmath 2\pi f t}, \\ \text{with} & H_{\mathcal{F}}(f) & = & \int_{-\infty}^{+\infty} h(\tau)\, e^{-\jmath 2\pi f \tau}\, d\tau \end{array}$$

The proportionality coefficient between $y(t)$ and $x(t)$, $H_{\mathcal{F}}(f)$, is the Fourier transform of $h(t)$.

More generally, if $y(t)$ is the response to an excitation $x(t)$ of a stationary linear system, then $Y_{\mathcal{F}}(f) = H_{\mathcal{F}}(f)\, X_{\mathcal{F}}(f)$, where $X_{\mathcal{F}}(f)$ and $Y_{\mathcal{F}}(f)$ are the Fourier transforms of signals $x(t)$ and $y(t)$. This relationship demonstrates the possibility of conditioning measurement signals using linear systems, which behave, in the frequency domain, as a mask applied to the input signal in order to preserve information (taking $H_{\mathcal{F}}(f) \approx 1$) and to remove disturbances (using $H_{\mathcal{F}}(f) \approx 0$). These systems therefore allow us to separate information from noise, if their frequency locations are different.

3.3.1. *Analytical case studies*

EXERCISE 3.12.– An *ideal* low-pass filter, which leaves the signal totally unaffected in the passband, and completely eliminates everything outside of it,

$$H(f) = \begin{cases} 1, & |f| \leq f_c \\ 0, & |f| > f_c \end{cases}$$

cannot be created, in reality. With electronic components, it is only possible to create filters with a rational transfer function. Butterworth low-pass filters[3] are a group of technically feasible filters whose behavior is fairly close to that of ideal low-pass filters. The square modulus of their frequency response is equal to:

$$|H(f)|^2 = \frac{1}{1 + (f/f_c)^{2N}}$$

3 Stephen Butterworth (1880–1958), an English engineer, first worked at the British National Physical Laboratory (NPL) and then at the Admiralty Research Laboratory until 1945, [BUT 30].

This family therefore uses the following two values as parameters:

– The cutoff frequency f_c ($f_c \in I\!R_+^*$).

– The N-order of the filter ($N \in I\!N^*$).

Hence, in order to solve the problem of designing a low-pass filter, we find the values of these two parameters that satisfy specifications compiled on the basis of the problem at hand, as follows:

– For all frequencies between 0 and a given frequency f_1, the frequency response must not have a square modulus less than a given value $1 - \epsilon_1$:

$$\forall f,\ |f| \leq f_1 \Longrightarrow 1 - \epsilon_1 \leq |H(f)|^2 \leq 1,\ \text{where } 0 < \epsilon_1 \ll 1 \qquad [3.1]$$

This frequency f_1 is the upper bound of the passband of the filter, and ϵ_1 is linked to the maximum deformation of the signal in that band.

– For all frequencies above a given frequency f_2, ($f_2 > f_1$), the filter's frequency response must have a square modulus less than a given value ϵ_2:

$$\forall f,\ |f| \geq f_2 \Longrightarrow 0 < |H(f)|^2 \leq \epsilon_2,\ \text{where } 0 < \epsilon_2 < 1 - \epsilon_1 \qquad [3.2]$$

This frequency f_2 is the lower bound of the stopband of the filter, and ϵ_2 is linked to the minimum attenuation of the signal in the stopband.

The interval $[f_1, f_2]$ is the transition band of the filter, within which range the designer imposes no constraints.

The objective of this exercise is to show how to deduce the values of N and f_c from the four values of f_1, f_2, ϵ_1 and ϵ_2 given by the technical specifications.

1) On the basis of equation [3.1], show that f_c and N must satisfy the inequality:

$$\left(\frac{f_c}{f_1}\right)^{2N} > \frac{1-\epsilon_1}{\epsilon_1} \qquad [3.3]$$

2) On the basis of equation [3.2], show that f_c and N must satisfy the inequality:

$$\left(\frac{f_2}{f_c}\right)^{2N} > \frac{1-\epsilon_2}{\epsilon_2} \qquad [3.4]$$

3) Deduce from this the inequality:

$$N > \frac{\log_{10}\left(\sqrt{\frac{(1-\epsilon_1)(1-\epsilon_2)}{\epsilon_1\epsilon_2}}\right)}{\log_{10}(\frac{f_2}{f_1})}$$

Which value of N should be chosen? How does the value of N evolve when $\frac{f_2}{f_1}$ and/or ϵ_1 and ϵ_2 vary?

4) When the value of N is chosen, show that relations [3.3] and [3.4] necessitate that f_c be in an interval $[f_{\text{cmin}}, f_{\text{cmax}}]$. Show that the frequency $\sqrt{f_{\text{cmin}} f_{\text{cmax}}}$ lies within this interval, and can therefore be chosen. What does this value become if $\epsilon_1 = \epsilon_2$?

5) Deduce from this the parameters of a Butterworth filter, which satisfies the following technical specifications:

$$f_1 = 30 \text{ Hz}, \epsilon_1 = 0.01, \quad f_2 = 400 \text{ Hz}, \epsilon_2 = 0.03$$

What happens if f_1 is multiplied by 10? What happens if ϵ_1 and ϵ_2 are divided by 10?

EXERCISE 3.13.– The objective of this exercise is to show how it is possible to create a bandpass filter with adjustable central frequency, using two low-pass filters. This technique is known as bandpass filtering by quadrature modulation[4].

The principle is illustrated in Figure 3.5. Two signals $x_1(t)$ and $x_2(t)$, obtained by multiplication of the input signal $x(t)$ by $\sqrt{2}\cos(\omega_0 t)$ and $\sqrt{2}\sin(\omega_0 t)$ (where $\omega_0 = 2\pi f_0$), excite two identical filters with frequency response $H(f)$. The outputs of these filters are multiplied by the same signals, so that $y_1(t) = z_1(t)\sqrt{2}\cos(\omega_0 t)$ and $y_2(t) = z_2(t)\sqrt{2}\sin(\omega_0 t)$, and then added to construct the output signal $y(t) = y_1(t) + y_2(t)$.

1) With the supposition that $x(t) = X\,e^{\jmath 2\pi f t}$, calculate $x_1(t)$, $z_1(t)$ and $y_1(t)$. To do so, transform the term $\cos(\omega_0 t)$ into a sum of two complex exponentials. Verify that $x_1(t)$ is the sum of two complex exponentials of frequencies $f + f_0$ and $f - f_0$, and that $y_1(t)$ is the sum of three complex exponentials. What are their frequencies?

2) Calculate $x_2(t)$, $z_2(t)$ and $y_2(t)$ in the same way. Finally, deduce that $y(t) = G(f)\,X\,e^{\jmath 2\pi f t}$, where $G(f) = H(f - f_0) + H(f + f_0)$.

3) Using the fact that a signal can be written in the form $x(t) = \int_{-\infty}^{+\infty} X(f)\,e^{\jmath 2\pi f t}\,df$, deduce the expression of $y(t)$.

4) If $H(f)$ is an ideal low-pass filter,

$$H(f) = \begin{cases} 1 & \text{if} \quad |f| < f_c \\ 0 & \text{else} \end{cases},$$

where $f_c \ll f_0$, summarily represent $G(f)$.

5) The two low-pass filters have real impulse responses iff $\forall f \in I\!R$, $H(f)^* = H(-f)$. This is a necessary condition for

4 For further details about this structure, see [HAZ 00].

them to actually be feasible. Show that $G(f)$ also has a real impulse response.

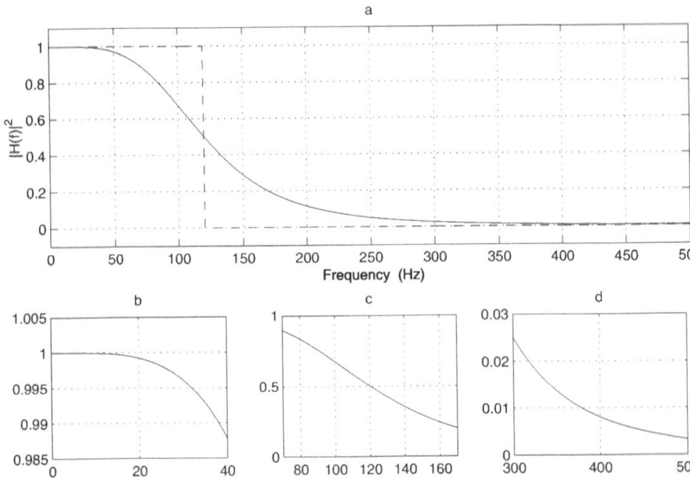

Figure 3.4. *Example of a problem of designing a low-pass filter: the filter (see Figure 3.4(a)) must not deform the components of the signal less than 30 Hz ($\epsilon_1 = 0.01$, $f_1 = 30$ Hz), and must eliminate the components of the signal greater than 400 Hz ($\epsilon_2 = 0.03$, $f_2 = 400$ Hz). The solution chosen ($N = 2$, $f_c = 120$ Hz, see Figure 3.4(c)), which reasonably closely approximates the ideal low-pass filter (dotted lines), does indeed satisfy the specifications in the passband (see Figure 3.4(b)) and in the stopband (see Figure 3.4(d))*

3.3.2. *Practical applications*

SoX allows easy filtering of signals using discrete-time equivalents of first- or second-order continuous-time filters. The command file below (`Traitement1.bat`) shows these possibilities:

```
rem filtrage d'un signal avec les commandes sox
rem lowpass, highpass, bandpass et bandreject
rem F. Auger, IUT Saint-Nazaire, dep. MP, jan. 2010
```

```
sox SonRigolo3.mp3 SonRigolo_lp1.mp3 lowpass  -1  500
sox SonRigolo3.mp3 SonRigolo_hp2.mp3 highpass -2 2000 0.8q
sox SonRigolo3.mp3 SonRigolo_bp2.mp3 bandpass     200 5q
sox SonRigolo3.mp3 SonRigolo_br2.mp3 bandreject
200 0.6q
sox SonRigolo_bp2.mp3 SonRigolo_bp2b.mp3 gain -20

rem pause
```

– The first command applies a first-order (-1) `lowpass` filter to signal `SonRigolo3.mp3` with a limiting frequency of 500 Hz. The output signal of the filter is stored in `SonRigolo_lp1.mp3`.

– The second command applies a second-order (-2) `highpass` filter to signal `SonRigolo.mp3` with a limiting frequency of $2,000$ Hz and a quality factor of $Q = 0.8$. The quality factor Q is equal to $1/(2m)$, where m is the damping coefficient. The output signal of the filter is stored in `SonRigolo_hp2.mp3`.

– The third command applies a second-order `bandpass` filter to signal `SonRigolo.mp3` with a central frequency of 200 Hz and factor $q = 5$. The output signal of the filter is stored in `SonRigolo_bp2.mp3`.

– The fourth command applies a second-order `bandreject` filter to signal `SonRigolo.mp3` with a central frequency of 200 Hz and factor $q = 0.6$. The output signal of the filter is stored in `SonRigolo_bp2.mp3`.

– All of the filters presented above have a unitary maximum gain. To amplify or attenuate the signal, we may use the `gain` action in SoX. The fifth command applies a gain of -20 dB (i.e. an attenuation of factor 10) to the signal obtained using the bandpass filter, and stores the result in `SonRigolo_bp2b.mp3`.

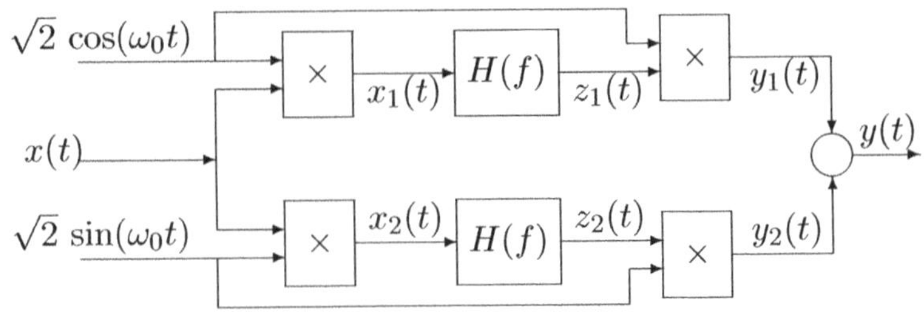

Figure 3.5. *Principle of a quadrature modulation filter, studied in Exercise 3.13. The rectangles containing × symbols are multipliers*

SOX also allows us to trace the modulus of the frequency response of these filters using the command `-]-plot gnuplot`. It generates a command file for the `gnuplot` curve tracing program[5]. SOX does not apply any processing operations to the input signal in this case, but the signal must still be identified in order for the program to obtain the sample rate. The command file shown below (`Traitement2.bat`), for example, produces the four frequency responses shown in Figure 3.6:

```
rem trace de la reponse frequentielle des filtres obtenus avec
rem les commandes sox lowpass, highpass, bandpass et bandreject
rem F. Auger, IUT Saint-Nazaire, dep. MP, jan. 2010

sox --plot gnuplot SonRigolo3.mp3 -n lowpass   -1   500       > Courbe1.plt
sox --plot gnuplot SonRigolo3.mp3 -n highpass -2 2000 0.8q > Courbe2.plt
sox --plot gnuplot SonRigolo3.mp3 -n bandpass       200 5q    > Courbe3.plt
sox --plot gnuplot SonRigolo3.mp3 -n bandreject     200 0.6q > Courbe4.plt

pause
```

The files generated by SOX, with extension `.plt`, are command files that enable `gnuplot` to produce graphics. By

5 `gnuplot` is available at http://www.gnuplot.info.

default, `gnuplot` calculates 100 values of the represented function. If this proves to be insufficient for a satisfactory graphical representation of the function, an instruction of type `set samples 300` may be used after the command:

```
set ylabel 'Amplitude Response (dB)'
```

Unfortunately, the `gnuplot` command file generated by SoX does not allow us to zoom in on specific zones of the graphical representation. To do this, we need to open the `gnuplot` command file (for example `Courbe1.plt`) using a text editor and remove the `[-35:25]` from command

```
plot [f=10:Fs/2] [-35:25] 20*log10(H(f))
```

which imposes upper and lower limits on the y `-axis`.

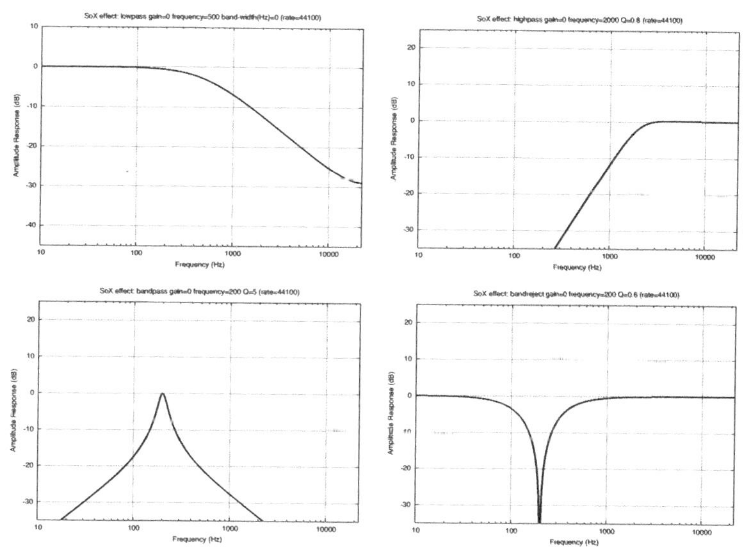

Figure 3.6. *Frequency responses obtained using* `lowpass`, `highpass`, `bandpass` *and* `bandreject` *commands with the addition of the* `-plot gnuplot` *command*

To obtain characteristic temporal responses for a filter, we may use and adapt the program below (TestFiltre.ch), which generates a signal containing an impulse, a step and a ramp, and then filter this signal using the appropriate SoX command. Figure 3.7 shows the signal produced by this program and the results obtained using command

```
sox TestFiltre.wav TestFiltre2.wav lowpass -1 10
```

```
//
Programme permettant de generer un fichier pour tester des filtres
//  F. Auger, juin 2013

#include <stdio.h>

int main()
{
 char   NomFichierDat[] = "TestFiltre.dat" ;
 FILE   *PointeurFichierDat ;
 int i, NbPoints=2000;
 double Fe=16000.0, Te=1.0/Fe, t=0.0, Amp=0.9;

 // generation du fichier .dat par une relecture du fichier texte
 PointeurFichierDat  =fopen(NomFichierDat,
"w");// ouverture en ecriture

 if ( PointeurFichierDat == NULL )
  {printf("Impossible d'ouvrir le fichier %s en ecriture\n",
          NomFichierDat);}
 else
  {
   // on precise la frequence d'echantillonnage du signal
   fprintf(PointeurFichierDat,"; Sample Rate %8.2f\n", Fe);

   // un seul signal = une seule voie = un seul canal
   fprintf(PointeurFichierDat,"; Channels 1\n");

   for(i=0;i<NbPoints;i=i+1)
    {fprintf(PointeurFichierDat, "%f %f \n", t, 0.0); t=t+Te;}

   // l'impulsion
   fprintf(PointeurFichierDat, "%f %f \n", t, Amp); t=t+Te;

   for(i=0;i<NbPoints;i=i+1)
    {fprintf(PointeurFichierDat, "%f %f \n", t, 0.0); t=t+Te;}

   // l'echelon
   for(i=0;i<NbPoints;i=i+1)
    {fprintf(PointeurFichierDat, "%f %f \n", t, Amp); t=t+Te;}
```

```
    for(i=0;i<NbPoints;i=i+1)
     {fprintf(PointeurFichierDat, "%f %f \n", t, 0.0); t=t+Te;}

    // la rampe
    for(i=0;i<NbPoints;i=i+1)
     {fprintf(PointeurFichierDat, "%f %f \n",
              t, Amp*(i+1)/NbPoints); t=t+Te;}

    for(i=0;i<NbPoints;i=i+1)
     {fprintf(PointeurFichierDat, "%f %f \n", t, 0.0); t=t+Te;}

    fclose(PointeurFichierDat);
  }
 return 0;
}
```

EXERCISE 3.14.– Note that the transfer function of a second-order bandpass filter with a central pulse of ω_0 and damping z is written as:

$$H(p) = \frac{2\,z\,\omega_0\,p}{p^2 + 2\,z\,\omega_0\,p + \omega_0^2}$$

$$= \frac{\frac{\omega_0}{Q}\,p}{p^2 + \frac{\omega_0}{Q}\,p + \omega_0^2} = \frac{1}{1 + Q\left(\frac{p}{\omega_0} + \frac{\omega_0}{p}\right)}$$

$$\text{with} \quad \omega_0 = 2\pi\,f_0 \quad \text{and} \quad Q = \frac{1}{2\,z} = \frac{\omega_0}{\Delta\omega} = \frac{f_0}{\Delta f}$$

The aim of this exercise is to study the variation of the actual central frequency and the bandwidth Δf of a second-order bandpass filter as a function of its quality factor Q and the desired central frequency f_0. To do this, we trace the frequency response of the bandpass filters obtained using different values of the quality factor Q (for example $Q = 2^{n/2}$, for n from -2 to 3) and of f_0 (for example $f_0 = 300\,2^k$ for k from 0 to 5), and then measure the actual central frequency and the bandwidth Δf for each response in order to compare them with theoretical values. To do this, we may use two nested loops as shown in the command file below (passe_bande.bat):

```
rem F. Auger, IUT Saint-Nazaire, dep. MP, mars 2012
rem %%i et %%j sont deux variables
rem le symbole ^ en fin de ligne indique
rem que la commande continue sur la ligne suivante

for %%f in (300,600,1200,2400,4800,9600) do ^
for %%Q in (0.5, 0.707, 1, 1.414, 2, 2.8) do ^
sox --plot gnuplot SonRigolo3.mp3 -n bandpass
    %%f %%Qq > bode_%%f_%%Q.plt

rem pause
```

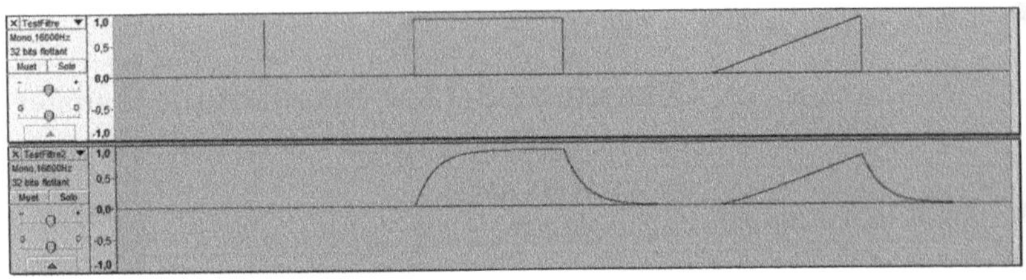

Figure 3.7. *Signal obtained using program* `TestFiltre.ch` *and response to the signal from a first-order "analog" low-pass filter with a limiting frequency of 10 Hz*

EXERCISE 3.15.– Filter the first signal obtained in Exercise 1.2 using a band-reject filter with a central frequency of 500 Hz and a rejection band of 100 Hz. Measure the amplitude of the successive sinusoids to deduce experimental values for the rejection bandwidth and the modulus of the frequency response of the filter, which should be added to the theoretical curve.

EXERCISE 3.16.– Command

```
sox SonRigolo.mp3 SonRigolo_lp2.mp3 lowpass -2 500 3q
```

may be used to apply the time-discrete equivalent of a second-order analog low-pass filter, with a static gain of 1, a natural frequency of $f_0 = 500$ Hz and quality factor $Q = 3$ to signal `SonRigolo.mp3`. The output signal of this filter is stored in file `SonRigolo_lp2.mp3`. Note that the transfer function of a second-order low-pass filter with a natural frequency of f_0 and a quality factor of Q is written as:

$$H(p) = \frac{\omega_0^2}{p^2 + \frac{\omega_0}{Q}\,p + \omega_0^2} = \frac{\omega_0^2}{p^2 + 2\,z\,\omega_0\,p + \omega_0^2}$$

where $\omega_0 = 2\pi\, f_0$ and $Q = \frac{1}{2\,z}$, i.e. $z = \frac{1}{2\,Q}$. The modulus curve of the frequency response of this filter (expressed in decibels) may be obtained using the command[6]

```
sox --plot gnuplot SonRigolo.mp3 -n lowpass -2 500 3q > lowpass2.plt
```

This generates a file that may be used with `gnuplot` to obtain a plot of the type shown in Figure 3.8.

If Q is greater than $\sqrt{2}/2$, this filter presents resonance (i.e. a maximum higher than 1 has a non-null frequency) of frequency:

$$f_r = f_0\sqrt{1 - \frac{1}{2Q^2}} = f_0\sqrt{\frac{2Q^2 - 1}{2Q^2}}$$

6 The curve obtained with this command is constructed using 100 values from the frequency response. If this is not sufficient to obtain a satisfactory graphical representation, we may modify the `.plt` file by adding an instruction of type `set samples 300` after the command `set ylabel 'Amplitude Response (dB)'`. This instruction allows us to produce a curve using 300 values from the frequency response. Note that in order to zoom in on the graphic, we must simply remove `"[-35:25]"` from the instruction `plot [f=10:Fs/2] [-35:25] 20*log10(H(f))`.

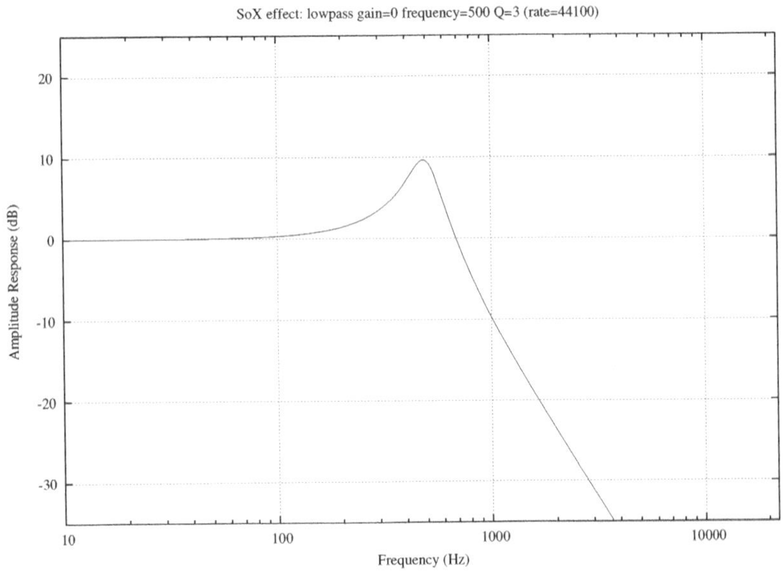

Figure 3.8. *Frequency response of a second-order low-pass filter obtained using the lowpass command followed by the* `-plot gnuplot` *command*

The frequency response modulus for this frequency is equal to:

$$G_r = |H(\jmath 2\pi f_r)| = \frac{2Q^2}{\sqrt{4Q^2 - 1}} = Q\sqrt{1 + \frac{1}{4Q^2 - 1}}$$

Note that G_r is independent of f_0. The limiting frequency at -3 dB is equal to:

$$f_c = f_0 \sqrt{1 - \frac{1}{2Q^2} + \sqrt{2 - \frac{1}{Q^2} + \frac{1}{4Q^4}}}$$

Calculate $H(\jmath\omega_0)$. What are the limits of f_r, G_r and f_c when Q tends toward infinity? What are the limits of f_r, G_r and f_c when Q tends toward $\sqrt{2}/2$? Write a command file to use SOX in order to plot the frequency response of second-order low-pass filters obtained with two values of Q (2 and 20) and two values of f_0 (1000 and 10,000 Hz). Using these four curves, measure the resonance frequency f_r, the resonance gain G_r, the gain at frequency f_0 and the limiting frequency f_c. Compare these results to the theoretical values and discuss your findings.

3.4. Digital filtering

3.4.1. *Analytical case studies*

EXERCISE 3.17.– To eliminate the high-frequency components of a discrete-time signal $x[n]$, we can envisage using averaging filters. Let $y_1[n]$ and $y_2[n]$ be the signals defined by

$$\begin{aligned} y_1[n] &= \frac{1}{3}\{x[n] + x[n-1] + x[n-2]\} \\ y_2[n] &= \frac{1}{4}\{x[n] + x[n-1] + x[n-2] + x[n-3]\} \end{aligned}$$

S_1 is the system with input $x[n]$ and output $y_1[n]$ and S_2 the system with input $x[n]$ and output $y_2[n]$.

1) Calculate the static gains of these two systems and analyze the results obtained.

2) Calculate the Nyquist gains of these two systems and analyze the results obtained.

3) Express $y_1[n-1]$ as a function of $x[n-1]$, $x[n-2]$ and $x[n-3]$, and $y_2[n-1]$ as a function of $x[n-1]$, $x[n-2]$, $x[n-3]$ and $x[n-4]$. Deduce the expressions of $y_1[n] - y_1[n-1]$ and

$y_2[n] - y_2[n-1]$. Deduce from this the frequency responses (or complex gains, or transfer functions) $H_1(\lambda)$ and $H_2(\lambda)$ for these two systems.

4) Calculate $H_1(\frac{1}{3})$ and $H_2(\frac{1}{4})$ and analyze the results obtained.

5) Using an expansion limited to the 3rd third order of the sine function $\sin(x) \approx x - x^3/6$, calculate an approximate value for the cutoff frequency of these two filters, and draw a conclusion from this.

EXERCISE 3.18.– On the basis of a discrete-time signal $x[n]$ obtained by sampling with the period T_s of a signal $x(t)$, we construct a second signal $y[n]$ using the relation[7]:

$$y[n] = \alpha\, y[n-1] + \frac{1+\alpha}{2}\left(x[n] - x[n-1]\right),$$

where α is a real parameter between -1 and 1.

1) Does this relation define a linear system? Does it define a steady-state system? Justify your answers.

2) Calculate the static gain and the Nyquist gain. Analyze the results obtained.

3) For $\alpha = 1/3$, calculate the response of the system with a unit step. We will carefully explain why, in this case, $y[-1] = 0$. Analyze the results obtained, and make the link with the previous question.

4) Calculate the frequency response of this filter, and recalculate the results for Exercise 3.2. Using SoX, plot the

7 Such a system is called a DC blocking filter.

frequency response of this filter for different values of α, and compare the results obtained.

5) Using, for instance, the expansions limited to the first order of the exponential and the function $1/(1+x)$,

$$e^x = 1 + x + x\,\epsilon(x) \quad \text{and} \quad \frac{1}{1+x} = 1 - x + x\,\epsilon(x),$$

deduce that the slope of this transfer function at the origin is proportional to $\frac{1+\alpha}{2\,(1-\alpha)}$. What can we conclude from this? Deduce an approximation of the cutoff frequency for this filter. Is the approximation obtained a minorant or a majorant of the true value?

6) Figures 3.9 and 3.10 show the modulus and phase of this frequency response for multiple values of α. Characterize these responses and find the results from the previous question. Justify the use of this filter to eliminate the continuing undesirable components from certain signals.

7) This filter is excited by a sinusoidal filter of frequency λ. Plot the filter's response for $\lambda = 1/3$ and for different values of λ. Perform a significant number of tests, and analyze the results obtained. Find the link with the previous questions.

EXERCISE 3.19.– Consider the discrete-time system that, based on a sampled signal $x[n]$, constructs a new signal $y[n]$ defined by the relation:

$$\begin{aligned} y[n] &= b\,x[n] + b\,x[n-1] + a\,x[n-2] - a\,x[n-4] \\ &\quad - b\,x[n-5] - b\,x[n-6], \end{aligned}$$

where a and b are two real parameters.

1) Calculate the static gain and the Nyquist gain of this system. Interpret the results obtained.

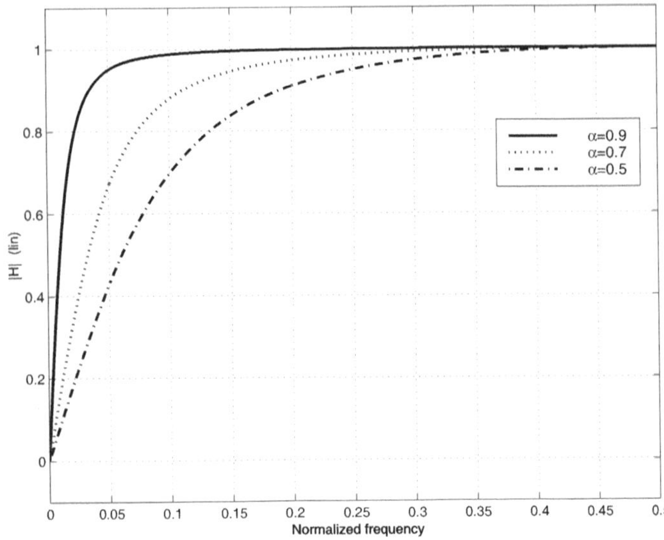

Figure 3.9. *Modulus of the frequential response of the filter studied in Exercise 3.18*

2) Using the superposition theorem, deduce the system's response to a signal that fluctuates slightly around a constant x_0, alternately taking the values $x_0 + u_x$ and $x_0 - u_x$. For example, we could take $x[n] = x_0 + u_x\,(-1)^n$.

3) Show that this system's frequency response can take the form

$$H(\lambda) = 2\jmath\, \{a\, \sin(2\pi\lambda) + b\, \sin(4\pi\lambda) + b\, \sin(6\pi\lambda)\}\; e^{-\jmath 6\pi\lambda}$$

4) Using the expansion limited to the third order of $\sin(x)$, calculate a and b so that $H(\lambda) \approx \jmath 2\pi\lambda\, e^{-\jmath 6\pi\lambda}$. We then obtain a filter that gives an approximation of the derivative of the signal $x[n]$, with a delay of three samples, so $y[n] \approx T_s\, \frac{dx}{dt}((n - 3)\, T_s)$.

EXERCISE 3.20.– We are now interested in a discrete-time system that, on the basis of a sampled signal $x[n]$, constructs

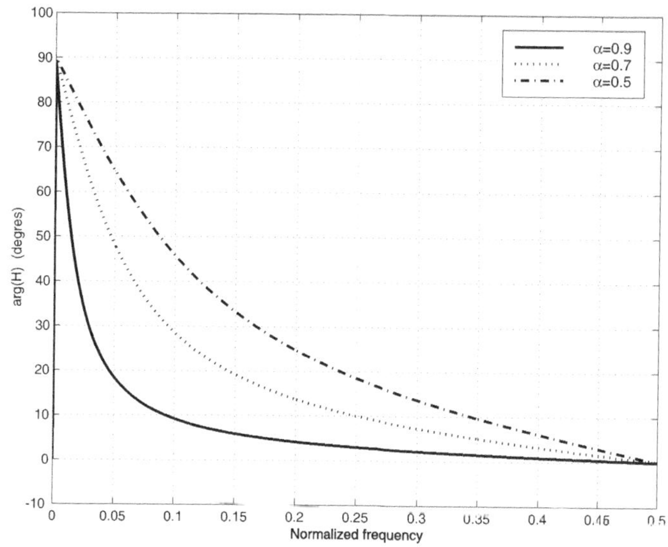

Figure 3.10. *Phase of the frequency response of the filter studied in Exercise 3.18*

a new signal $y[n]$ by the relation[8]:

$$y[n] = \frac{1}{16}\left(x[n] + 2\,x[n-1] + 3\,x[n-2] + 4\,x[n-3]\right.$$
$$\left. +3\,x[n-4] + 2\,x[n-5] + x[n-6]\right)$$

1) Calculate the first terms of this filter's impulse response. Does this filter have a finite or infinite impulse response?

2) Calculate the static gain of this filter, i.e. the ratio between the output signal and the input signal when these two signals are constant: $x[n] = x[n-1] = x[n-2] = X$ and $y[n] = y[n-1] = y[n-2] = Y$. Interpret the results obtained.

8 See [CIR 05].

3) Calculate the Nyquist gain of this filter, i.e. the ratio between the output signal and the input signal when these two signals are alternate series: $x[n] = (-1)^n X = -x[n-1] = x[n-2]$ and $y[n] = (-1)^n Y = -y[n-1] = y[n-2]$. Interpret the results obtained.

4) Calculate the frequency response H(λ) of this filter, i.e. the ratio between the output signal and the input signal when these two signals are of the form $x[n] = X\, z^n$ and $y[n] = Y\, z^n$, where $z = e^{\jmath 2\pi\lambda}$.

5) From this, deduce that this frequency response can be written in the form

$$H(\lambda) = \frac{1}{16}\left(1 + z^{-1} + z^{-2} + z^{-3}\right)^2 = \left(\frac{1 - z^{-4}}{4(1 - z^{-1})}\right)^2$$

Using the relations of symmetrization of the type $1 - x^{-2} = x^{-1}\left(x - x^{-1}\right)$, deduce that

$$H(\lambda) = e^{-\jmath 6\pi\lambda}\left(\frac{\sin(4\pi\lambda)}{4\,\sin(\pi\lambda)}\right)^2$$

6) Using a second-order Taylor approximation around zero,

$$\frac{\sin(K\,x)}{K\,\sin(x)} \approx 1 - \frac{K^2 - 1}{6} x^2,$$

calculate an approximate value of the cutoff frequency for this filter. Analyze the quality of that approximation using Figure 3.11.

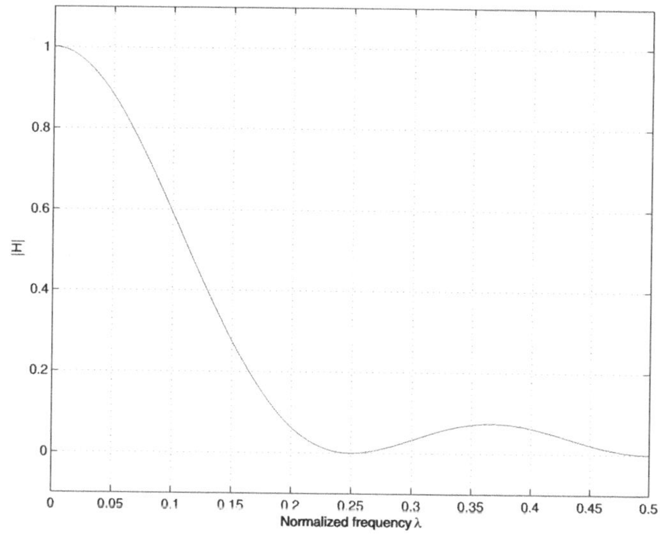

Figure 3.11. *Representation of the modulus (linear scale) of the frequency response of the filter studied in Exercise 3.20, as a function of the normalized frequency* $\lambda = f\,T_e = f/F_s$

EXERCISE 3.21.– We are interested in a discrete-time system that, based on a sampled signal $x[n]$, constructs a new signal $y[n]$ by way of the relation[9]

$$y[n] = a\,x[n] + b\,x[n-1] + 2\,c\,x[n-2] + b\,x[n-3] + a\,x[n-4]$$

1) Calculate the static gain of this filter, i.e. the ratio between the output signal and the input signal when these two signals are constant: $x[n] = x[n-1] = x[n-2] = X$ and $y[n] = y[n-1] = y[n-2] = Y$. Interpret the results obtained.

2) Calculate the Nyquist gain of this filter, i.e. the ratio between the output signal and the input signal when these

9 See [HAM 73].

two signals are alternate series: $x[n] = (-1)^n X = -x[n-1] = x[n-2]$ and $y[n] = (-1)^n Y = -y[n-1] = y[n-2]$. Interpret the results obtained.

3) Calculate the frequency response $H(\lambda)$ of this filter, i.e. the ratio between the output signal and the input signal when these two signals are of the form $x[n] = X\, z^n$ and $y[n] = Y\, z^n$, where $z = e^{\jmath 2\pi\lambda}$.

4) Using symmetrization relations of the type $1 + x^{-2} = x^{-1}\,(x + x^{-1})$, deduce that this frequency response can be written in the form:

$$\begin{aligned} H(\lambda) &= A(\lambda)\, e^{-\jmath 4\pi\lambda} \\ \text{where } A(\lambda) &= 2a\, \cos(4\pi\lambda) + 2b\, \cos(2\pi\lambda) + 2c \end{aligned}$$

5) To calculate a, b and c, we impose that $A(0) = 1$ (static gain equal to 1), $A(0.5) = 0$ (Nyquist gain equal to 0) and $\frac{d^2A}{d\lambda^2}(0) = 0$. Deduce the values of a, b and c. What type of filter is thus created, and what is its cutoff frequency?

EXERCISE 3.22.– We are interested in a discrete-time system which, based on a sampled signal $x[n]$, constructs a new signal $y[n]$ by the relation

$$y[n] = K\,\left(x[n] - 2\alpha\, x[n-1] + 2\alpha\, x[n-3] - x[n-4]\right)$$

1) Calculate the first terms of this filter's response to an impulse at time zero. Does this filter have a finite or infinite impulse response? Does this impulse response verify a property of symmetry?

2) Calculate the static gain of this filter, i.e. the ratio between the output signal and the input signal when these two signals are constant: $x[n] = x[n-1] = x[n-2] = X$ and $y[n] = y[n-1] = y[n-2] = Y$. Interpret the results obtained.

3) Calculate the Nyquist gain of this filter, i.e. the ratio between the output signal and the input signal when these two signals are alternate series: $x[n] = (-1)^n X = -x[n-1] = x[n-2]$ and $y[n] = (-1)^n Y = -y[n-1] = y[n-2]$. Interpret the results obtained.

4) Calculate the frequency response $H(\lambda)$ of this filter, i.e. the ratio between the output signal and the input signal when these two signals are of the form $x[n] = X z^n$ and $y[n] = Y z^n$, where $z = e^{\jmath 2\pi\lambda}$.

5) Using symmetrization relations of the type $1 + x^{-2} = x^{-1}(x + x^{-1})$, deduce that this frequency response can be written in the form:

$$\begin{aligned} H(\lambda) &= \jmath A(\lambda)\, e^{-\jmath 4\pi\lambda} \\ \text{where } A(\lambda) &= 4K \sin(2\pi\lambda)\,(\cos(2\pi\lambda) - \alpha) \end{aligned}$$

6) To calculate α and K, we impose that $A(\lambda_0) = 1$ and $\frac{dA}{d\lambda}(\lambda_0) = 0$. Thus, we obtain a bandpass filter with central frequency λ_0. Deduce the values of α and K.

7) Figure 3.12 shows the modulus of the frequency response obtained for $\lambda_0 = 0.2$. What is the passband of the filter obtained?

EXERCISE 3.23.– A system generates a signal $y[n]$ on the basis of another signal $x[n]$ by a relation of the form:

$$\begin{aligned} y[n] &= h_0\, x[n] + h_1\, x[n-1] + h_2\, x[n-2] \\ &\quad + h_1\, x[n-3] + h_0\, x[n-4] \end{aligned}$$

1) Which relation must be verified by the coefficients h_0, h_1 and h_2 so that $y[n]$ is always null when $x[n]$ is a constant signal, of the type $x[n] = x_0$?

2) Which additional relation must be satisfied by the coefficients h_0, h_1 and h_2 so that $y[n]$ is always null when $x[n]$ is a linear signal, of the type $x[n] = v_0\, T_s\, n$?

3) Which additional relation must be satisfied by the coefficients h_0, h_1 and h_2 so that $y[n]$ is always equal to $\gamma_0\, T_s^2$ when $x[n]$ is a parabolic signal, of the type $x[n] = \frac{\gamma_0\, T_s^2}{2}\, n^2$?

4) Which additional relation must be satisfied by the coefficients h_0, h_1 and h_2 so that $y[n]$ is equal to $j_0\, T_s^3\, n$ when $x[n]$ is equal to $x[n] = \frac{j_0\, T_s^3}{6}\, n^3$?

5) Deduce from this the values of h_0, h_1 and h_2. What, then, is the value of the signal $y[n]$ when $x[n] = x_0 + v_0\, T_s\, n + \frac{\gamma_0\, T_s^2}{2}\, n^2 + \frac{j_0\, T_s^3}{6}\, n^3$?

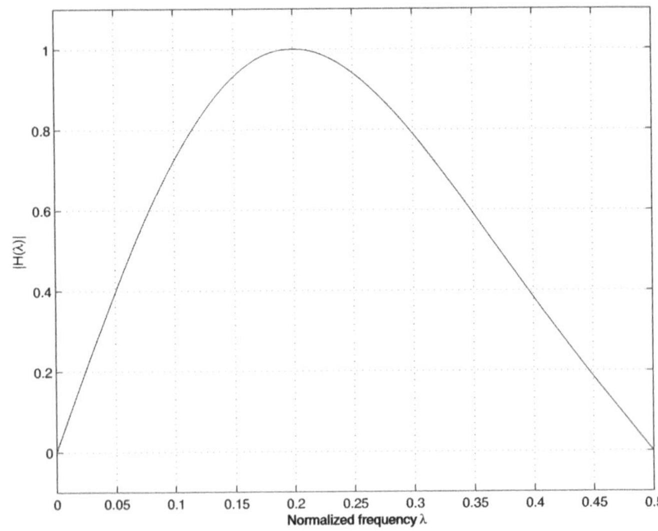

Figure 3.12. *Representation of the modulus (linear scale) of the frequency response of the filter studied in Exercise 3.22, as a function of the normalized frequency* $\lambda = f\, T_s = f/F_s$

3.4.2. *Practical applications*

SOX also allows us to carry out discrete-time filtering using the commands `fir` and `biquad`. The command file below (`Traitement3.bat`) illustrates the possibilities of these two commands:

```
rem filtrage a temps discret a l'aide des commandes sox fir et biquad
rem F. Auger, IUT Saint-Nazaire, dep. MP, jan. 2010

sox SonRigolo.mp3 SonRigolo_fir1.mp3 fir 0.1 0.2 0.4 0.3

sox SonRigolo.mp3 SonRigolo_fir2.mp3 fir CoeffsFIR.txt

sox SonRigolo.mp3 SonRigolo_biq.mp3 biquad 0.6 0.2 0.4 1 -1.5 0.6

sox --plot gnuplot SonRigolo.mp3 -n fir 0.2 0.2 0.2 0.2 0.2 > CourbeFir.plt

rem pause
```

– The first command uses the signal $x[n]$ contained in the file `SonRigolo.mp3` to create a new signal $y[n]$, obtained using a third-order finite impulse response (*FIR*) filter of equation:

$$y[n] = 0.1\,x[n] + 0.2\,x[n-1] + 0.4\,x[n-2] + 0.3\,x[n-3]$$

– The second command uses a second `FIR` filter for which the coefficients are contained in file `CoeffsFIR.txt`. In this file, the coefficients are separated by spaces or placed onto separate lines (return/enter key). Lines that start with the character # are comment lines. An example of content from the `CoeffsFIR.txt` file is shown below:

```
# filtre RIF du 4eme ordre
# F. Auger, jan 2010
0.35
0.62
0.51
0.62
0.35
```

– The third command uses the signal $x[n]$ contained in the file `SonRigolo.mp3` to create a new signal $y[n]$, obtained using a second-order infinite impulse response (`IIR`) filter of equation:

$$\begin{aligned} y[n] &= 0.6\,x[n] + 0.2\,x[n-1] + 0.4\,x[n-2] \\ &\quad +1.5\,y[n-1] - 0.6\,y[n-2] \end{aligned}$$

– The fourth command traces the modulus of the frequency response of a fourth-order `FIR` filter of equation:

$$y[n] = \frac{1}{5}\left(x[n] + x[n-1] + x[n-2] + x[n-3] + x[n-4]\right)$$

The result is shown in Figure 3.13.

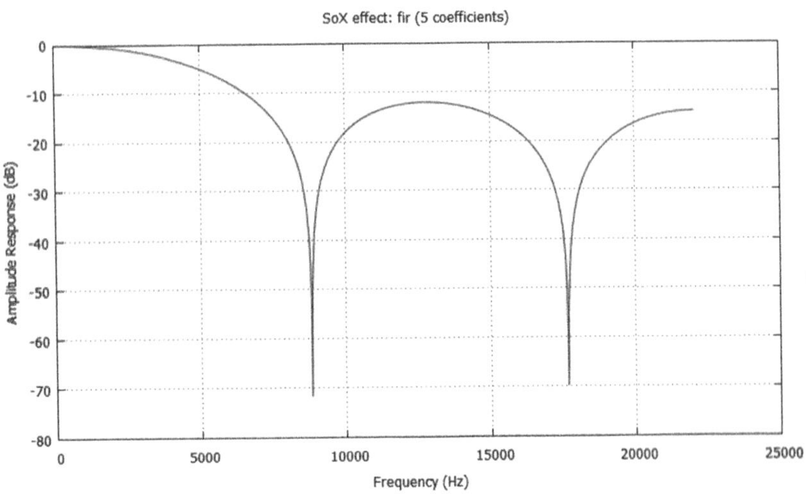

Figure 3.13. *Frequency response of the* `FIR` *filter obtained using the fourth command from file* `Traitement3.bat`.

EXERCISE 3.24.– Filter a signal made up of a sequence of sinusoids with suitable frequencies using an averaging filter of length 10, with equation:

$$y[n] = \frac{1}{10}\left(\sum_{k=0}^{9} x[n-k]\right)$$

Deduce experimental measurements of the modulus of the frequency response of this filter at the successive frequencies of the signal and compare them to theoretical values. Verify, experimentally, that the frequency response of this filter is equal to zero at frequency $F_s/10$, and determine, experimentally, its cutoff frequency.

EXERCISE 3.25.– The command `sinc` allows us to filter a signal using a specific type of digital `FIR` filters. The command file below (`Partiel2010c.bat`) illustrates its possible uses:

```
rem Filtrage avec la commande sinc de SoX
rem F. Auger, IUT Saint-Nazaire, dep. MP, mars 2010

sox InputFile.mp3 OutputFile1.mp3 sinc -3500
sox InputFile.mp3 OutputFile2.mp3 sinc -3500 -n 43
sox InputFile.mp3 OutputFile3.mp3 sinc 4000

sox --plot gnuplot InputFile.mp3 -n sinc -12000 -n 151
> FreqResp.plt

rem pause
```

– The first command applies a low-pass filter with a cutoff frequency (at -6 dB) of $3,500$ Hz to the signal contained in file `InputFile.mp3`. The resulting output signal is stored in file `OutputFile1.mp3`.

– The second command operates in the same way, specifying that `InputFile.mp3` should be filtered using a filter of order 42 (the output of the filter is then calculated from 43 values of the input signal). If this is not specified by the user, as in the first case, SOX selects the order of the filter itself.

– The third command applies a high-pass filter with a cutoff frequency (at -6 dB) of $4,000$ Hz.

– The final command plots the modulus of the frequency response of a low-pass filter of order 150 with a cutoff frequency (at -6 dB) of 12 kHz.

Design a signal including a sequence of sinusoids with frequencies ranging from 200 to 1,400 Hz, by steps of 50 Hz. Each sinusoid should have a duration of 1.5 s. Next, filter this signal using a low-pass filter of order 215 with a cutoff frequency (at -6 dB) of 800 Hz. In what form should the relationship between signal $y[n]$, obtained as output from this filter, and signal $x[n]$, the input, be written? Measure the amplitudes of the sinusoids obtained as input and output using this filter and deduce experimental values of the modulus of the frequency response. Compare these values to the theoretical curve plotted using the program.

We wish to filter a signal in order to keep only those components located between 310 and $1,240$ Hz, with the exception of components between 490 and 510 Hz. Propose a solution using filters of this type and apply it to the signal generated at the beginning of this exercise to check that we obtain the desired frequency response.

Bibliography

[AUG 99] AUGER F., *Introduction à la théorie du signal et de l'information: cours et exercices*, Technip, Paris, 1999.

[BEL 02] BELLANGER M., *Traitement numérique du signal*, 9th ed., collection Sciences Sup, Dunod, Paris, 2002.

[BEN 02] BENIDIR M., *Théorie et traitement du signal, vol. 1 of Représentation des signaux et des systèmes*, collection Sciences Sup, Dunod, Paris, 2002.

[BIN 13] BINET G., *Traitement numérique du signal – Signaux et systèmes discrets*, collection Technosup, Ellipses, Paris, 2013.

[BUT 30] BUTTERWORTH S, "On the theory of filter amplifiers," *Experimental Wireless and the Wireless Engineer*, vol. 7, pp. 536–541, 1930.

[CIR 05] CIRRUS LOGIC, "CS5358: low power channel decimation filter", available at http://ww.cirrus.com, October 2005.

[COT 09] COTTET F., *Traitement des signaux et acquisition de données*, collection Sciences Sup, Dunod, Paris, 2009.

[HAM 73] HAMMING R.W., *Numerical Methods for Scientists and Engineers*, 2nd ed., Dover Publication Inc., p. 593, 1973.

[HAZ 00] HAZRA S.N., "Bandpass filtering using quadrature modulation", *IEEE Trans. on Circuits and Systems II*, vol. 46, no. 12, pp. 1516–1518, 2000.

[LYO 04] LYONS R., *Understanding Digital Signal Processing*, 2nd ed., Prentice Hall PTR, Upper Saddle River, NJ, 2004.

[LYO 12] LYONS R., *Streamlining Digital Processing: A Trick of the Trade Guidebook*, Wiley/IEEE Press, Hoboken, NJ, 2012.

[MAR 05] MARTIN J.N., *Débuter en traitement numérique du signal*, collection Technosup, Ellipses, Paris, 2005.

[MUI 11] MUIRA K., *Haut-parleur directionnel à ultrasons*, Elektor, pp. 64–67, 2011.

[OPP 10] OPPENHEIM A., SCHAFER R., *Discrete-Time Signal Processing*, 3rd ed., Prentice Hall/Pearson, Upper Saddle River, NJ, 2010.

[TIS 04] TISSERAND E., PAUTEX J.F., SCHWEITZER P., *Analyse et traitement des signaux: méthodes et applications au son et à l'image*, Dunod, Paris, 2004.

Index

Printed and bound by CPI Group (UK) Ltd, Croydon, CR0 4YY
13/09/2026
14961081-0006